Lecture Notes in Physics

W0235200

The Editorial Policy for Edited Volumes

The series *Lecture Notes in Physics* (LNP), founded in 1969, reports new developments in physics research and teaching - quickly, informally but with a high degree of quality. Manuscripts to be considered for publication are topical volumes consisting of a limited number of contributions, carefully edited and closely related to each other. Each contribution should contain at least partly original and previously unpublished material, be written in a clear, pedagogical style and aimed at a broader readership, especially graduate students and nonspecialist researchers wishing to familiarize themselves with the topic concerned. For this reason, traditional proceedings cannot be considered for this series though volumes to appear in this series are often based on material presented at conferences, workshops and schools.

Acceptance

A project can only be accepted tentatively for publication, by both the editorial board and the publisher, following thorough examination of the material submitted. The book proposal sent to the publisher should consist at least of a preliminary table of contents outlining the structure of the book together with abstracts of all contributions to be included. Final acceptance is issued by the series editor in charge, in consultation with the publisher, only after receiving the complete manuscript. Final acceptance, possibly requiring minor corrections, usually follows the tentative acceptance unless the final manuscript differs significantly from expectations (project outline). In particular, the series editors are entitled to reject individual contributions if they do not meet the high quality standards of this series. The final manuscript must be ready to print, and should include both an informative introduction and a sufficiently detailed subject index.

Contractual Aspects

Publication in LNP is free of charge. There is no formal contract, no royalties are paid, and no bulk orders are required, although special discounts are offered in this case. The volume editors receive jointly 30 free copies for their personal use and are entitled, as are the contributing authors, to purchase Springer books at a reduced rate. The publisher secures the copyright for each volume. As a rule, no reprints of individual contributions can be supplied.

Manuscript Submission

The manuscript in its final and approved version must be submitted in ready to print form. The corresponding electronic source files are also required for the production process, in particular the online version. Technical assistance in compiling the final manuscript can be provided by the publisher's production editor(s), especially with regard to the publisher's own LaTeX macro package which has been specially designed for this series.

LNP Homepage (springerlink.com)

On the LNP homepage you will find:
−The LNP online archive. It contains the full texts (PDF) of all volumes published since 2000. Abstracts, table of contents and prefaces are accessible free of charge to everyone. Information about the availability of printed volumes can be obtained.
−The subscription information. The online archive is free of charge to all subscribers of the printed volumes.
−The editorial contacts, with respect to both scientific and technical matters.
−The author's / editor's instructions.

U.-G. Meißner W. Plessas (Eds.)

Lectures on Flavor Physics

 Springer

Editors

Ulf-G. Meißner
Universität Bonn
Helmholtz-Institut
für Strahlen- und Kernphysik
Nussallee 14-16
53115 Bonn, Germany

Willibald Plessas
Universität Graz
Institut für Theoretische Physik
Universitätsplatz 5
8010 Graz, Austria

Supported by the Österreichische Bundesministerium für Bildung, Wissenschaft und Kultur, Vienna, Austria

---

U.-G. Meißner W. Plessas (Eds.), *Lectures on Flavor Physics*, Lect. Notes Phys. **629** (Springer, Berlin Heidelberg 2004), DOI 10.1007/b98411

Library of Congress Control Number: 2004107013

ISSN 0075-8450
ISBN 978-3-662-14521-0   ISBN 978-3-540-44457-2 (eBook)
DOI 10.1007/978-3-540-44457-2

springeronline.com

© Springer-Verlag Berlin Heidelberg 2004

Originally published by Springer-Verlag Berlin Heidelberg New York in 2004
Softcover reprint of the hardcover 1st edition 2004

The use of general descriptive names, registered names, trademarks, etc. in this publication does not imply, even in the absence of a specific statement, that such names are exempt from the relevant protective laws and regulations and therefore free for general use.

Typesetting: Camera-ready by the authors/editor
Data conversion: PTP-Berlin Protago-TeX-Production GmbH
Cover design: *design & production*, Heidelberg

Printed on acid-free paper
54/3141/ts - 5 4 3 2 1 0

Preface

This volume contains the written versions of some selected lectures delivered at the "41. Internationale Universitätswochen für Theoretische Physik" in Schladming, Austria. The 41st "Schladming Winter School" took place during the period February 22nd–28th, 2003. The theme of the School was "Flavor Physics".

Flavor physics is one of the hot topics in contemporary elementary particle physics, because it relates to fundamental questions like the origin of masses, the size and strength of CP violation, or the oscillations between various neutrino species. One thus explores the Standard Model as well as its possible extensions and related phenomena in astrophysics and cosmology. The lectures collected in this volume deal with important (theoretical) developments at various length scales. At low energies and for light quarks, one is able to analyze the strong interactions in terms of an effective field theory, as described in Jürg Gasser's lectures on light-quark dynamics. The interrelation between precisely calculable electroweak and the much more difficult strong interactions comes into play in precision QED observables, like the anomalous magnetic moment of the muon as discussed in the lectures by Marc Knecht. In fact, the presently available precision data might already give a glimpse at physics beyond the Standard Model.

The issue of CP violation in kaon and B-meson systems is addressed in the lectures by Andrzej Buras. Only with the advent of the B-factories, CP violation in the Standard Model could be measured beyond the kaon system, and one is now testing the unitarity of the CKM matrix and tries to understand the fundamental mechanism underlying CP violation. The richness of this field is underlined by the lectures of Matthias Neubert, who presents a new scheme to systematically calculate strong-interaction effects in certain B-decays and also dwells on the unitarity triangle. Last but not least, with the recent measurements of neutrino oscillations at Super-Kamiokande, SNO, and Kamland, the review on the foundations of the various forms of flavor transitions in vacuum and in media by Walter Grimus is very timely.

At the School, we had further lectures on the experimental status of B-decays (by D. Hitlin) as well as neutrino physics (by G. Drexlin), on lattice calculations of flavor-physics matrix elements (by G. Martinelli), and an introduction to supersymmetry (by H. Dreiner). They are not included here, but their essential content can be found from other sources in the literature.

Here, we should like to express our sincere gratitude to the lecturers for all their efforts in preparing and presenting their lectures. We are especially grateful to those colleagues who managed to find time to write up their lectures. We thank also the main sponsors of the School, the Austrian Federal Ministry for Education, Science, and Culture as well as the Government of Styria, for providing financial support. In addition, we acknowledge the contributions from the University of Graz and the valuable organisational and technical assistance by the town of Schladming, Ricoh Austria, and Hornig Graz. Furthermore, we are grateful to our secretaries, S. Fuchs and E. Monschein, a number of graduate students from our institute, and, last but not least, our colleagues from the organizing committee for their valuable assistance in preparing and running the school.

Bonn and Graz,   *Ulf-G. Meißner*
December 2003   *Willibald Plessas*

Contents

5   The Angles α, β and γ from B Decays 114
   5.1   Preliminaries 114
   5.2   Classification of Elementary Processes 114
   5.3   Neutral B Decays into CP Eigenstates 115
   5.4   Decays to CP Non-eigenstates 118
   5.5   U–Spin Strategies 121
   5.6   Constraints for γ from $B \to \pi K$ 121
6    $K^+ \to \pi^+ \nu \bar{\nu}$ and $K_{\mathrm{L}} \to \pi^0 \nu \bar{\nu}$ 124
   6.1   Branching Ratios 124
   6.2   Unitarity Triangle and $\sin 2\beta$ from $K \to \pi \nu \bar{\nu}$ 126
   6.3   Concluding Remarks 126
7   Minimal Flavour Violation Models 127
   7.1   Preliminaries 127
   7.2   Universal Unitarity Triangle 128
   7.3   Models with Universal Extra Dimensions 128
8   Outlook 128
   8.1   Phase 1 (2003-2007) 128
   8.2   Phase 2 (2007-2009) 129
   8.3   Phase 3 (2009-2013) 130

Heavy-Quark Physics
Matthias Neubert 137
1   Introduction 137
2   Statement of the Factorization Formula 140
   2.1   The Idea of Factorization 140
   2.2   The Factorization Formula 141
3   Arguments in Favor of Factorization 142
   3.1   Preliminaries and Power Counting 142
   3.2   Non-leptonic Decay Amplitudes 145
   3.3   Remarks on Final-State Interactions 147
4   Power-Suppressed Contributions 149
   4.1   Interactions with the Spectator Quark 149
   4.2   Annihilation Topologies 150
   4.3   Non-leading Fock States 151
5   Limitations of the Factorization Approach 153
   5.1   Several Small Parameters 153
   5.2   Power Corrections Enhanced by Small Quark Masses 154
6   QCD Factorization for Charmless Decays 155
7   Testing Factorization in $B \to \pi K, \pi\pi$ Decays 157
8   Establishing CP Violation in the Bottom Sector 158
9   Mixing-Independent Construction
   of the Unitarity Triangle 162
10   Outlook 165

Neutrino Physics – Theory

List of Contributors

Jürg Gasser
Institut für Theoretische Physik
Universität Bern
Sidlerstraße 5
3012 Bern, Switzerland
gasser@itp.unibe.ch

Marc Knecht
Centre de Physique Théorique
CNRS Luminy
Case 907
13288 Marseille Cedex 9, France
knecht@cpt.univ-mrs.fr

Andrzej J. Buras
Technical University Munich
Physics Department
85748 Garching, Germany
Andrzej_Buras@ph.tum.de

Matthias Neubert
Institute for High-Energy Phenomenology
Newman Laboratory for Elementary-Particle Physics
Cornell University
Ithaca, NY 14853, U.S.A.
neubert@lepp.cornell.edu

Walter Grimus
Institute for Theoretical Physics
University of Vienna
Boltzmanngasse 5
1090 Vienna, Austria
walter.grimus@univie.ac.at

Light-Quark Dynamics

Jürg Gasser

Institut für Theoretische Physik, Universität Bern, Sidlerstraße 5, 3012 Bern, Switzerland, `gasser@itp.unibe.ch`

1 Introduction

At low energies $E \ll M_W$, the interactions of leptons and hadrons are described by QCD + QED up to corrections of order (E/M_W). If we disregard the electromagnetic interactions, we are left with QCD that contains only a few parameters: the renormalization group invariant scale Λ and the running quark masses $m_u, m_d, m_s \ldots$. The quark masses m_u, m_d and m_s are small on a typical hadronic scale like the mass of the rho or of the proton. It makes therefore sense to consider the limit where these masses are set equal to zero (chiral limit). The remaining quarks $c, b, \ldots$ are not light: although one may of course study the theoretical limit in which these masses also vanish, it does not seem to be possible to recover the actual mass values by an expansion around that limiting case. At low energies, a better approximation is obtained if the quarks $c, b, \ldots$ are instead treated as infinitely heavy. In this limit, the degrees of freedom associated with these quarks freeze and may be ignored in the effective low energy theory.

In the chiral limit, QCD contains therefore only one parameter, the scale Λ. The mass of the proton is a pure number multiplying Λ, and likewise for all the other states of the theory – the numbers $M_\rho/M_p, M_\Delta/M_p, \ldots$ are determined in a parameter free manner. In this sense, the chiral limit of QCD may be called a theory without any adjustable parameters: QCD is of course unable to predict the value of M_p in GeV units, but it determines all dimensionless hadronic quantities in a parameter free manner. The elastic cross section for pp scattering e.g. is some fixed function of the variables s/M_p^2 and t/M_p^2 , multiplying the square of the Compton wavelength of the proton.

It is unfortunately very difficult to really *calculate* masses, cross sections and decay amplitudes in this beautiful theory, because the lagrangian of QCD is formulated in terms of quark and gluon fields which do not create asymptotically observed particles. Several methods have therefore been devised in the past to cope with this problem in different regimes of the energy scale:

i) Processes at high energies. At high energies, the effective coupling constant α_{QCD} becomes small, and conventional perturbation theory in α_{QCD} is applicable.

ii) Lattice calculations. This is the only method known today which leads directly from the QCD lagrangian to the mass spectrum, decay matrix elements, scattering lengths etc. On the other hand, the CPU time needed for

J. Gasser, Light-Quark Dynamics, Lect. Notes Phys. **629**, 1–35 (2004)
http://www.springerlink.com/   © Springer-Verlag Berlin Heidelberg 2004

full fledged QCD calculations is enormous, and I believe that one may still have to wait a long time before this program achieves the accuracy one is aiming at in the framework of effective field theory.

iii) Chiral perturbation theory (ChPT). This method exploits the symmetry of the QCD lagrangian and its ground state: one solves in a perturbative manner the constraints imposed by chiral symmetry and unitarity by expanding the Green functions in powers of the external momenta and of the quark masses m_u, m_d and m_s. To illustrate the idea, consider the process $\pi^+(p_1)\pi^-(p_2) \to \pi^0(p_3)\pi^0(p_4)$. Chiral symmetry implies that the corresponding scattering amplitude has the following form near threshold,

$$T = \frac{M_\pi^2 - s}{F_\pi^2} + O(p^4) \; ; \; s = (p_1 + p_2)^2 , \tag{1}$$

where $F_\pi = 92.4$ MeV is the pion decay constant, and M_π denotes the pion mass. This result is due to Weinberg [1], who used current algebra and PCAC to analyse the Ward identities for the four-point functions of the axial currents. It displays the first order term in a systematic expansion of the scattering amplitude in powers of momenta and of quark masses. This term algebraically dominates the remainder, denoted by the symbol $O(p^4)$, for sufficiently small energies and thus provides an accurate parameterization of the full amplitude near threshold. As one goes away from threshold, the higher order terms come into play. We will see in the following that ChPT is a method that allows one to determine these corrections in a systematic manner.

ChPT is a particular example of an *effective field theory* (EFT). The method is in use since about 20 years, and it was therefore not possible to provide a detailed review in my lectures – for a recent comprehensive introduction to ChPT, I refer the reader to [2]. Instead, I discussed a few basic principles and applications, in the hope that students become interested in this fascinating topic and continue with their own studies and research projects.

The article is organized as follows. In Sect. 2, the flavor symmetries of QCD are discussed, and their Nambu-Goldstone realization explained. In Sect. 3, the Goldstone theorem is stated and illustrated with the free scalar field, with the linear sigma model ($L\sigma M$) and with QCD. In addition, the interaction of the Goldstone bosons at low energy is investigated. Section 4 contains a discussion of the effective field theory of the $L\sigma M$ and of QCD at low energy. In Sect. 5 are illustrated some calculations with these EFT, and Sect. 6 contains a detailed discussion of the elastic $\pi\pi$ scattering amplitude in this framework. In Sect. 7, it is shown how Roy equations may be used to determine low-energy constants that appear in the calculation of the $\pi\pi$ scattering amplitude. A short outlook on other topics is given in Sect. 8.

2   QCD with Two Flavours

In this section, I discuss the flavour symmetries of QCD.

2.1   Symmetry of the Lagrangian

The lagrangian of QCD is

$$\mathcal{L} = -\frac{1}{2g^2}\langle G_{\mu\nu}G^{\mu\nu}\rangle_c + \mathcal{L}_{ud} ,\tag{2}$$

where

$$\mathcal{L}_{ud} = \bar{u}\,D\!\!\!/\,u + \bar{d}\,D\!\!\!/\,d - m_u\bar{u}u - m_d\bar{d}d$$

$$= (\bar{u}\ \bar{d})\begin{pmatrix} D\!\!\!/ - m_u & 0 \\ 0 & D\!\!\!/ - m_d \end{pmatrix}\begin{pmatrix} u \\ d \end{pmatrix} \ ; \ D\!\!\!/ = i\gamma^u(\partial_\mu - iG_\mu) .$$

$G_{\mu\nu}$ denotes the field strength associated with the gluon field G_μ, and $\langle A\rangle_c$ stands for the color trace of the matrix A.

It is useful to introduce left- and right-handed spinors,

$$u_L = \frac{1}{2}(1 - \gamma_5)u \ , \ u_R = \frac{1}{2}(1 + \gamma_5)u \ ,$$

$$\mathcal{L}_{ud} = (\bar{u}_L\ \bar{d}_L)\begin{pmatrix} D\!\!\!/ & 0 \\ 0 & D\!\!\!/ \end{pmatrix}\begin{pmatrix} u_L \\ d_L \end{pmatrix} - (\bar{u}_L\ \bar{d}_L)\begin{pmatrix} m_u & 0 \\ 0 & m_d \end{pmatrix}\begin{pmatrix} u_R \\ d_R \end{pmatrix} + L \leftrightarrow R .$$

QCD makes sense for any value of the quark masses. For $m_u = m_d = 0$, the lagrangian (2) is invariant under $U(2)$ rotations of the left- and right-handed fields,

$$\begin{pmatrix} u_I \\ d_I \end{pmatrix} \Rightarrow V_I\begin{pmatrix} u_I \\ d_I \end{pmatrix} \ ; \ V_I \in U(2) \ , \ I = L, R .\tag{3}$$

In other words, gluon interactions do not change the helicity of the quarks, see Fig. 1. On the other hand, the terms proportional to the quark masses are not invariant under the transformations (3), see Fig. 2.

According to the theorem of E. Noether, there is one conserved current for each continuous parameter in the symmetry group. As the group $U(2)$ has four real parameters, one expects eight conserved currents. However, due to quantum effects, one of these currents is not conserved, as a result of which there are only seven conserved currents in the limit of vanishing quark masses,

$$L_\mu^a = \bar{q}_L\gamma_\mu\frac{\tau^a}{2}q_L \ , \ R_\mu^a = \bar{q}_R\gamma_\mu\frac{\tau^a}{2}q_R \ ; \ a = 1,2,3$$

$$V_\mu = \bar{q}\gamma_\mu q \ ; \ q = \begin{pmatrix} u \\ d \end{pmatrix} .\tag{4}$$

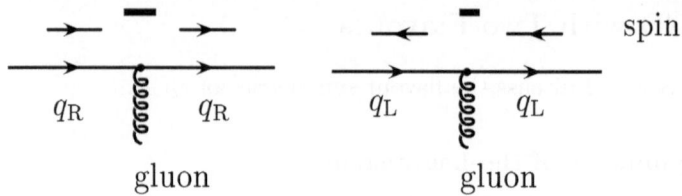

Fig. 1. Gluon interactions do not change the helicity of the quarks

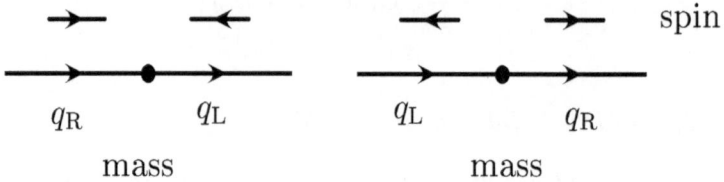

Fig. 2. The mass terms change the helicity of the quarks

The world at

$$m_u = m_d = 0$$

is called *the chiral limit of QCD*, and the above statements are summarized as:
*In the chiral limit, $\mathcal{L}_{QCD}$ is symmetric under global $SU(2)_L \times SU(2)_R \times U(1)_V$
transformations. The corresponding 7 Noether currents (4) are conserved.*

2.2   Symmetry of the Ground State

It is useful to introduce in addition the vector and axial currents

$$V^{\mu a} = \bar{q}\gamma^\mu \frac{\tau^a}{2} q = L^{\mu a} + R^{\mu a} \, ,$$

$$A^{\mu a} = \bar{q}\gamma^\mu \gamma_5 \frac{\tau^a}{2} q = R^{\mu a} - L^{\mu a} \; ; \; a = 1, 2, 3 \, .$$

The corresponding 6 axial and vector charges $Q^a_{A,V}$ are conserved and commute with the hamiltonian $H_0 = H_{QCD}|_{m_u = m_d = 0}$,

$$[H_0, Q^a_V] = [H_0, Q^a_A] = 0 \; ; \; a = 1, 2, 3 \, . \tag{5}$$

Consider now eigenstates of H_0,

$$H_0 |\psi\rangle = E|\psi\rangle \, .$$

Then the states $Q^a_A |\psi\rangle$ and $Q^a_V |\psi\rangle$ have the same energy E, but carry opposite parity. On the other hand, there is no trace [3] of such a symmetry in nature. The resolution of the paradox has been provided by Nambu and

Lasinio back in 1960 [4]: whereas the vacuum is annihilated by the vector charges, it is not invariant under the action of the axial charges,

$$Q_V^a|0\rangle = 0 \ , \ Q_A^a|0\rangle \neq 0 \ . \tag{6}$$

There are two important consequences of this assumption:

i) The spectrum of H_0 contains three massless, pseudoscalar particles (Goldstone bosons (GB); Goldstone [5]). We will see more of this in the following section.
ii) The axial charges Q_A^a, acting on any state in the Hilbert space, generate Goldstone bosons,

$$Q_A^a|\psi\rangle = |\psi, G_1, \ldots, G_N, \ldots\rangle \ .$$

These are not one-particle states, and are therefore not listed in PDG, and there is therefore no contradiction anymore.

A theory with (5), (6) is called *spontaneously broken*: the symmetry of the hamiltonian is not the same as the symmetry of the ground state.

Where are the three massless, pseudoscalar states? The three pions $\pi^\pm, \pi^0$ are the lightest hadrons. They are not massless, because the quark masses are not zero in the real world [6]:

$$m_u \simeq 5\,\text{MeV} \ ,$$
$$m_d \simeq 9\,\text{MeV} \ . \tag{7}$$

In the following, we assume that the flavour symmetry of QCD is spontaneously broken to the diagonal subgroup,

$$\boxed{SU(2)_L \times SU(2)_R \rightarrow SU(2)_V}$$

and work out the consequences for the *interactions* between the Goldstone bosons.

Remark: Vafa and Witten [7] have shown that – modulo highly plausible assumptions – the vector symmetry $SU(2)_V$ is not spontaneously broken.

2.3   A Remark on Isospin Symmetry

Even if the quarks are not massless, the QCD lagrangian has a residual symmetry at $m_u = m_d$: it is invariant under the transformations (3) with $V_R = V_L \in SU(2)$. This symmetry is called *isospin symmetry*. We know from textbooks that isospin symmetry violations in the strong interactions are small[1]. On the other hand, according to (7), one has

$$m_d/m_u \simeq 1.8 \ . \tag{8}$$

---

[1] There are two sources of isospin violations: those due to electromagnetic interactions, and those due to the difference in the up and down quark masses

How can then isospin be a good symmetry if the quark masses differ so much? Consider the neutral and the charged pions: is it so that their masses differ by

$$(M^2_{\pi^+} - M^2_{\pi^0})/M^2_{\pi^0} \simeq \frac{m_d - m_u}{m_d + m_u} \simeq 0.3 ?$$

The answer is no: one has

$$M^2_{\pi^+} = (m_u + m_d)B + \cdots ,$$
$$M^2_{\pi^0} = (m_u + m_d)B + \cdots ,$$

where the ellipses denote higher order terms in the quark mass expansion. The neutral and the charged pion have the same leading term, the quark mass difference shows up only in the quadratic piece,

$$M^2_{\pi^+} - M^2_{\pi^0} = O[(m_u - m_d)^2] .$$

The perturbation due to the quark masses can be written as

$$m_u \bar{u}u + m_d \bar{d}d = \frac{1}{2}(m_u + m_d)(\bar{u}u + \bar{d}d)$$
$$+ \frac{1}{2}(m_u - m_d)(\bar{u}u - \bar{d}d) .$$

Isospin is a good symmetry, not because $(m_d - m_u)/(m_d + m_u)$ is small, but because the matrix elements of the operator $\frac{1}{2}(m_d - m_u)(\bar{u}u - \bar{d}d)$ are small with respect to the hadron masses. The bulk part in the pion mass difference is generated by electromagnetic interactions.

3   Goldstone Bosons

In this section, I discuss the Goldstone theorem, illustrate it with several examples and consider the interaction of Goldstone bosons at low energy.

3.1   The Goldstone Theorem

We consider a quantum field theory which has the following properties:

i) There is a conserved current (i.e., an object that transforms as a four-vector under proper Lorentz transformations),

$$A_\mu(x) ; \quad \partial^\mu A_\mu = 0.$$

ii) There is an operator $\Phi(x)$ such that

$$\langle 0|[Q, \Phi]|0\rangle \neq 0 ; \quad Q = \int d^3x A_0(x^0, \vec{x}). \tag{9}$$

Then the Goldstone theorem [5] applies:

1. There exists a massless particle in the theory,

$$|\pi(\mathbf{p})\rangle \, , \; p^2 = 0 \, .$$

2. The current A_μ couples to the massless state,

$$\langle 0|A_\mu(0)|\pi(\mathbf{p})\rangle = ip_\mu F \neq 0 \, .$$

From the condition (9), it is seen that the charge Q does not annihilate the vacuum.

3.2 The Free Scalar Field

We begin with a very simple example, the free, massless scalar field. The lagrangian is given by

$$\mathcal{L}_0 = \frac{1}{2}\partial_\mu\phi\partial^\mu\phi \, .$$

For the current A_μ, we take

$$A_\mu = \partial_\mu\phi \, .$$

This current is conserved, because ϕ is a free field. Consider now $\Phi = \phi$. From the canonical commutation relations, it follows that the condition (9) is satisfied. Therefore, the Goldstone theorem applies. Indeed, we can easily check directly:

• ϕ generates massless states ,

and

• $\langle 0|A_\mu(0)|\pi\rangle = -ip_\mu \neq 0 \, .$

3.3 The Linear Sigma Model

We consider the linear sigma model ($L\sigma M$), because it allows one to illustrate many features of effective field theories. At the same time, it serves as a model with spontaneous symmetry breaking. The lagrangian is

$$\mathcal{L}_\sigma = \frac{1}{2}\partial_\mu\vec{\phi} \cdot \partial^\mu\vec{\phi} - \frac{g}{4}(\vec{\phi}^2 - v^2)^2 \, , \tag{10}$$

where $\vec{\phi} = (\phi^0, \phi^1, \phi^2, \phi^3)$ denotes four real fields, and $\vec{\phi}^2 = \phi^k\phi^k$ [repeated indices are summed over in the absence of a summation symbol]. In the following, we assume that

$$v^2 > 0 \, ,$$

and discuss

- the symmetry properties of $\mathcal{L}_\sigma$
- spontaneous symmetry breakdown
- Goldstone bosons
- quantization
- Goldstone boson scattering

Symmetry Properties. Here, we consider the classical theory and observe that $\mathcal{L}_\sigma$ is invariant under four-dimensional rotations of the vector $\vec{\phi}$,

$$\phi^i \to R^{ik}\phi^k \ , \ R \in O(4) \ .$$

The matrices R can be parametrized in terms of six real parameters. Let us consider infinitesimal rotations

$$R = 1 + \varepsilon + O(\varepsilon^2) \ .$$

Because R is an orthogonal matrix, ε is antisymmetric, $\varepsilon + \varepsilon^T = 0$. Every real and antisymmetric four by four matrix can be expanded in terms of six generators,

$$\varepsilon = \sum_{i=1}^{3}\left(c_i\varepsilon_V^i + d_i\varepsilon_A^i\right) \ ,$$

where c_i, d_i are 6 real parameters. The generators

$$\epsilon_A^1 \qquad\qquad \epsilon_A^2 \qquad\qquad \epsilon_A^3$$

$$\begin{pmatrix} 0 & -1 & 0 & 0 \\ 1 & 0 & 0 & 0 \\ 0 & 0 & 0 & 0 \\ 0 & 0 & 0 & 0 \end{pmatrix} \begin{pmatrix} 0 & 0 & -1 & 0 \\ 0 & 0 & 0 & 0 \\ 1 & 0 & 0 & 0 \\ 0 & 0 & 0 & 0 \end{pmatrix} \begin{pmatrix} 0 & 0 & 0 & -1 \\ 0 & 0 & 0 & 0 \\ 0 & 0 & 0 & 0 \\ 1 & 0 & 0 & 0 \end{pmatrix}$$

$$\begin{pmatrix} 0 & 0 & 0 & 0 \\ 0 & 0 & 0 & 0 \\ 0 & 0 & 0 & -1 \\ 0 & 0 & 1 & 0 \end{pmatrix} \begin{pmatrix} 0 & 0 & 0 & 0 \\ 0 & 0 & 0 & 1 \\ 0 & 0 & 0 & 0 \\ 0 & -1 & 0 & 0 \end{pmatrix} \begin{pmatrix} 0 & 0 & 0 & 0 \\ 0 & 0 & -1 & 0 \\ 0 & 1 & 0 & 0 \\ 0 & 0 & 0 & 0 \end{pmatrix}$$

$$\epsilon_V^1 \qquad\qquad \epsilon_V^2 \qquad\qquad \epsilon_V^3$$

satisfy the commutation relations
$$[\varepsilon_V^a, \varepsilon_V^b] = \varepsilon^{abc}\varepsilon_V^c \ ,$$
$$[\varepsilon_V^a, \varepsilon_A^b] = \varepsilon^{abc}\varepsilon_A^c \ ,$$
$$[\varepsilon_A^a, \varepsilon_A^b] = \varepsilon^{abc}\varepsilon_V^c \ ,$$
with $\varepsilon^{123} = 1$, cycl. The linear combinations

$$Q_L^a = \frac{1}{2}(\varepsilon_V^a - \varepsilon_A^a) \ , \ Q_R^a = \frac{1}{2}(\varepsilon_V^a + \varepsilon_A^a) \ ,$$

generate two commuting $SU(2)$ Lie-algebras (up to a factor i),

$$[Q_I^a, Q_I^b] = \varepsilon^{abc} Q_I^c \quad ; \quad I = L, R ,$$
$$[Q_L^a, Q_R^b] = 0 .$$

In other words, the lagrangian $\mathcal{L}_\sigma$ has a $SU(2)_L \times SU(2)_R$ symmetry. As a result of this, there are six conserved Noether currents, which we take to be

$$V_\mu^a = \varepsilon^{abc} \phi^b \partial_\mu \phi^c ,$$
$$A_\mu^a = -\phi^0 \overset{\leftrightarrow}{\partial_\mu} \phi^a \quad ; \quad a = 1, 2, 3 .$$

Spontaneous Symmetry Breaking. The potential $V = g(\vec{\phi}^{\,2} - v^2)^2/4$ is extremal at $\phi = 0$ and at $\vec{\phi}^{\,2} = v^2$. The latter configuration corresponds to a global minimum. The vector

$$\vec{\phi}_G = (v, \vec{0}) ,$$

which realizes this global minimum, is only invariant under the subgroup $H = O(3)$ (with generators ε_V^a), see Fig. 3 for the case where the symmetry group is $O(2)$. The number of generators that do not leave invariant $\vec{\phi}_G$ is $n_G - n_H = 3$, where

$$n_G : \text{ \# of parameters in O(4)} ,$$
$$n_H : \text{ \# of parameters in O(3)} .$$

Therefore, one expects three Goldstone bosons in the spectrum of the theory.

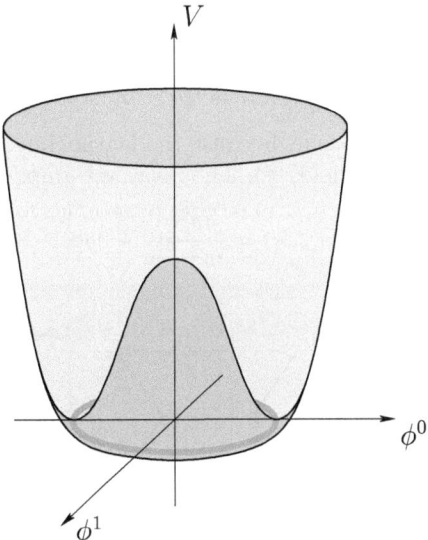

Fig. 3. Potential for $O(2)$

Goldstone Bosons. In order to identify the Goldstone bosons, we consider fluctuations around the configuration $\vec{\phi}_G$, and write

$$\vec{\phi} = (v + \varphi_0, \vec{\pi}) \ ; \ \vec{\pi} = (\pi^1, \pi^2, \pi^3) \ ,$$

where we have introduced the pion fields $\vec{\pi}$. In terms of the new fields, the lagrangian becomes

$$\mathcal{L}_\sigma = \frac{1}{2}[\partial_\mu\varphi_0\partial^\mu\varphi_0 - 2gv^2\varphi_0^2] + \frac{1}{2}\partial_\mu\vec{\pi}\cdot\partial^\mu\vec{\pi}$$
$$-gv\varphi_0\,(\varphi_0^2 + \vec{\pi}^{\,2}) - \frac{g}{4}(\varphi_0^2 + \vec{\pi}^{\,2})^2 \ . \tag{11}$$

The kinetic term shows that there is indeed one massive field φ_0, with mass $m = \sqrt{2g}v$, together with three massless fields $\vec{\pi}$.

Quantization. One evaluates Green functions generated by the lagrangian (11) in the standard manner. At tree level, one has one massive and three massless fields, as in the classical theory. Evaluating loops, one finds that φ_0 picks up a vacuum expectation value at order $\hbar$. One shifts this field again, such that the new field has a vanishing one-point Green function, and finds that the remaining three particles stay massless also in the loop expansion.

Goldstone Boson Scattering. Finally, we consider GB scattering at tree level, with the lagrangian (11). In particular, we consider the process

$$\pi^1(p_1)\pi^2(p_2) \to \pi^3(p_3)\pi^4(p_4) \ ,$$

where π^1 denotes the GB number one, etc. The relevant diagrams are displayed in Fig. 4. The scattering matrix element has the structure

$$T^{kl;ij} = \delta^{ij}\delta^{kl}A(s,t,u) + \text{cycl.} \ ,$$
$$s = (p_1 + p_2)^2, t = (p_3 - p_1)^2, u = (p_4 - p_1)^2 \ . \tag{12}$$

(The Kronecker symbols occur, because the lagrangian is invariant under an $O(3)$ rotation of the pion fields $\vec{\pi}$). The invariant amplitude is

$$A(s,t,u) = \frac{4g^2v^2}{m^2 - s} - 2g \ . \tag{13}$$

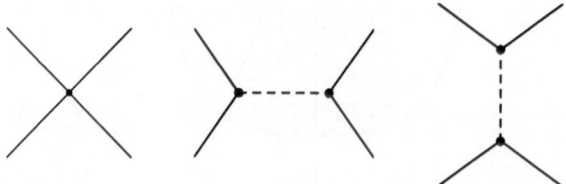

Fig. 4. Goldstone boson scattering with the lagrangian (11). *Solid (dashed)* lines stand for massless (massive) particles

At small momenta, the constant terms cancel out,

$$A(s,t,u) = \frac{s}{v^2} + O(s^2) . \tag{14}$$

In other words, the interaction between the GB vanishes at zero momenta. The constant v is related to the matrix elements of the axial current,

$$\langle 0|A_\mu^a(0)|\pi^b(p)\rangle = i\delta^{ab}vp_\mu .$$

3.4 QCD with Two Flavors

We now go back to the discussion of the Goldstone theorem in the framework of QCD. In Sect. 2, we noted that the three axial currents A_μ^a are conserved in the case of vanishing quark masses. For the field Φ in (9), we choose $\Phi = \bar{q}\gamma_5\tau^a q$ (three fields, one for each a). Applying canonical commutation relations, one finds that

$$[Q_A^a, \Phi^b] = -\delta^{ab}(\bar{u}u + \bar{d}d) \; ; \; a,b = 1,2,3 .$$

Provided that the vacuum expectation value of the quark bilinear is different from zero, the Goldstone theorem applies:

$$\langle 0|\bar{u}u|0\rangle \neq 0 \Rightarrow 3 \text{ Goldstone bosons} .$$

(The vacuum expectation value of $\bar{u}u$ is equal to the one of $\bar{d}d$ by isospin symmetry.) Lattice calculations support the conjecture that $\langle 0|\bar{u}u|0\rangle$ is different from zero. Later in these lectures, we shall see that data on K_{e4} decays do so as well.

How does one evaluate GB scattering in QCD? This is a very complicated affair: the QCD lagrangian contains quark and gluon fields, not pion fields. On the other hand, if QCD is spontaneously broken in the manner just discussed, the Goldstone theorem guarantees that the axial current can be used as an interpolating field for the pion [8]. Let us consider therefore the matrix element

$$G_\mu(p_4, p_3, p_1) - \langle \pi(p_3)\pi(p_4)\text{out}|A_\mu(0)|\pi(p_1)\rangle , \tag{15}$$

where I have suppressed all isospin indices. This matrix element has two parts: one, where the axial current generates a pion pole, and a second one, which is free from one-particle singularities, see Fig. 5. According to the LSZ reduction formula [9], the quantity G_μ has the structure

$$G_\mu = \frac{Fq_\mu}{q^2}T(p_3, p_4; p_1) + R_\mu , \tag{16}$$

where T denotes the elastic $\pi\pi$ scattering matrix element, and where F is the pion decay constant,

$$\langle 0|A_\mu(0)|\pi(p)\rangle = ip_\mu F .$$

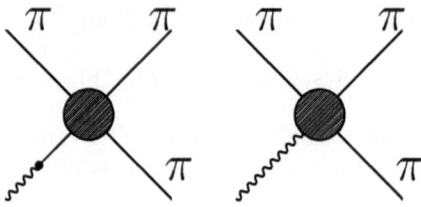

Fig. 5. Singular and non singular contributions to the matrix element (15) of the axial current. The *wavy line* denotes the axial current, the *solid lines* the pions

The remainder R_μ is non singular when $q^\mu = (p_3+p_4-p_1)^\mu$ is sent to zero. We contract both sides in (16) with q^μ. Because the axial current is conserved, the left-hand side vanishes, and therefore

$$FT(p_3, p_4; p_1) + q^\mu R_\mu = 0 .$$

One concludes that the scattering matrix element vanishes at $q^\mu = 0$ - this was easy to prove! We find again that the GB do not interact at vanishing momenta. One can even go further: as already mentioned in the introduction, Weinberg determined in 1966 [1] – using current algebra – the leading term of the $\pi\pi$ scattering amplitude in a systematic expansion of the momenta and of the pion mass.

4   Effective Field Theories

As we have just seen, it is possible to get a great deal of information about the interactions of GB in QCD without actually solving the theory. *Effective field theories* (EFT) provide the proper framework to perform detailed calculations in a systematic manner. EFT are valid in a restricted energy region, describing there an underlying theory that is valid on a wider energy scale. The situation for the Standard Model is illustrated in Fig. 6.

Effective field theories are in particular useful, when a full calculation is not yet possible, as is the case with the Standard Model at low energies. One sets up an EFT with the same symmetry properties – in case of the Standard Model, this effective theory is called *Chiral perturbation theory*. A second possibility for the use of EFT occurs in case that the calculations can be done in the underlying theory, but are very complicated. An example is provided by the calculation of bound states in the framework of QED, which can be performed, in principle, at any desired order by use of the Bethe-Salpeter equations. On the other hand, as Caswell and Lepage have shown [10], the use of EFT makes life very much simpler also in this case, because the electrons and positrons are moving very non-relativistically in the atoms, and a relativistic calculation is not really needed. Further applications concern the case where the underlying theory is not known – one builds the

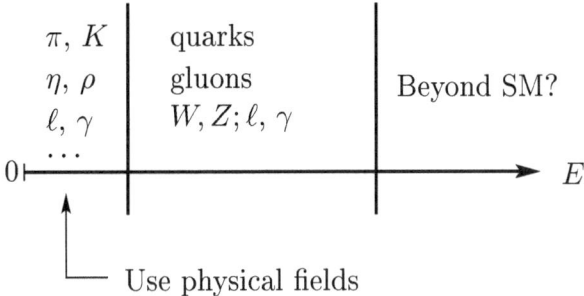

Fig. 6. The Standard Model at various energy scales. At low energy, the effective theory is formulated in terms of physical fields

effective theory in terms of the light fields and attempts to determine from experiment the unknown coupling constants that occur in there.

In the following, I discuss the low-energy EFT for the $L\sigma M$ and for QCD.

4.1   Linear Sigma Model at Low Energy

We have seen in the last section that the $O(4)$ version of the $L\sigma M$ in its broken phase develops three Goldstone bosons, which interact weekly at low energy (we have proven this at tree level only).

In the following, we construct an effective theory that contains only pions and their interactions - the sigma particle is removed from the theory. This EFT is constructed in such a manner that the Green functions with pion fields, evaluated in the $L\sigma M$ at low energies, are recovered by the EFT. How can this be achieved?

As a first step, one constructs an EFT which reproduces all *tree graphs* at low energies, to any order in the low-energy expansion: as is seen from the explicit expression (13), the scattering matrix element contains arbitrarily high powers in the momenta even at tree level. The procedure to construct the effective lagrangian is described in detail in the article by Nyffeler and Schenk [11] - I do not repeat the argument here and refer the interested reader instead to this work. The result can be written in various equivalent forms. Here, I use

$$\mathcal{L}_{\text{eff}}^{\sigma} = \mathcal{L}_2^{\sigma} + \mathcal{L}_4^{\sigma} + \cdots .$$

The lagrangians $\mathcal{L}_n^{\sigma}$ contain n derivatives of the pion fields. These derivatives become momenta in Fourier space: the effective lagrangian provides a momentum expansion of the amplitudes. Explicitly, one has for the leading term

$$\mathcal{L}_2^{\sigma} = \frac{v^2}{4} \langle \partial_\mu U \partial^\mu U^\dagger \rangle , \qquad (17)$$

where U is a 2×2 unitary matrix,

$$U = \sigma \cdot 1_{2\times 2} + \frac{i}{v}\tau^k\pi^k \quad ; \quad \sigma^2 + \frac{\vec{\pi}^2}{v^2} = 1 \, , \vec{\pi} = (\pi^1, \pi^2, \pi^3) \, .$$

The symbol $\langle A \rangle$ denotes the trace of the matrix A, and τ^k are the Pauli matrices. In order to illustrate the structure of this lagrangian, we expand it in terms of the pion fields:

$$\mathcal{L}_2^\sigma = \frac{1}{2}\partial_\mu\vec{\pi} \cdot \partial^\mu\vec{\pi} + \frac{1}{8v^2}\partial_\mu\vec{\pi}^2\partial^\mu\vec{\pi}^2 + O(\vec{\pi}^6) \, .$$

The lagrangians $\mathcal{L}_{4,6,8...}^\sigma$ have a similar structure [11].

Comments

- Only pions occur in the effective theory, the heavy particle has disappeared.
- It is easy to calculate the $\pi\pi$ scattering amplitude at tree level with $\mathcal{L}_2^\sigma$ - only one diagram remains to be calculated. The result is identical to the first term on the right-hand side of (14). The higher order terms in the expansion of the amplitude (13) are generated by the tree graphs of $\mathcal{L}_{4,6,..}^\sigma$.
- The mass of the sigma particle is given by $m = \sqrt{2g}v$. As a result of this, the interactions disappear formally in the large-mass limit.
- Where has the $O(4)$ symmetry of the $L\sigma M$ gone? $\mathcal{L}_2^\sigma$ is invariant under

$$U \to V_R U V_L^\dagger; \ V_{R,L} \in SU(2) \, .$$

Therefore, the effective theory has an $SU(2)_R \times SU(2)_L$ symmetry, as the original theory.
- How is the spontaneously broken symmetry realized?
$U_G = 1_{2\times 2}$ is the ground state. It is only invariant under

$$U_G \to V U_G V^\dagger \, .$$

Therefore, the theory is spontaneously broken,

$$SU(2)_R \times SU(2)_L \Rightarrow SU(2)_V \, ,$$

and generates three Goldstone bosons.

The lagrangian $\mathcal{L}_{\text{eff}}^\sigma$ reproduces the tree graphs of the $L\sigma M$ - how about loops? Indeed, one may calculate the scattering amplitudes in the framework of the $L\sigma M$ to any order in the loop expansion, perform a low-energy expansion of the result, and finally construct an effective theory that reproduces the result of this calculation, order by order in the low-energy expansion. The procedure is carried out to one loop in [12,11].

4.2   QCD at Low Energy

The effective theory of QCD is formulated in terms of asymptotic pion fields. As is the case for the $L\sigma M$, the effective theory consists of an infinite number of terms, with more and more derivatives [13,12]:

$$\mathcal{L}_{\text{eff}} = \mathcal{L}_2 + \mathcal{L}_4 + \mathcal{L}_6 + \cdots . \tag{18}$$

All that goes into the construction of $\mathcal{L}_{\text{eff}}$ are symmetry properties of QCD. The result for the leading term is

$$\mathcal{L}_2 = \frac{F^2}{4} \langle \partial_\mu U \partial^\mu U^\dagger + 2B\mathcal{M}(U + U^\dagger) \rangle , \tag{19}$$

where the field U is the same as above, and where

$$\mathcal{M} = \begin{pmatrix} m_u & 0 \\ 0 & m_d \end{pmatrix}$$

contains the quark masses m_u, m_d. A glance at (17) shows that, at $m_u = m_d = 0$, the leading order lagrangians in the $L\sigma M$ and in QCD agree, provided that one sets $v = F$. The leading term (19) contains the two constants F and B which are related to the pion decay constant and to the quark condensate, respectively [12],

$$\langle 0|A_\mu^a(0)|\pi^b(p)\rangle = ip_\mu F\delta^{ab} ,$$
$$\langle 0|\bar{u}u|0\rangle_{|m_u=m_d=0} = -F^2 B .$$

We have made a very big step: we have replaced the QCD lagrangian $\mathcal{L}_{\text{QCD}}$ by the effective lagrangian $\mathcal{L}_{\text{eff}}$. This transition may be visualized in terms of Feynman graphs: in Fig. 7 is displayed one of the infinitely many graphs that contribute to $\pi\pi$ scattering in QCD, and which are all equally important. On the other hand, in the effective theory, only the two graphs displayed in Fig. 8 contribute at leading order. In the figure, the symbol M^2 stands for the combination

$$M^2 = (m_u + m_d)B \tag{20}$$

which finally counts in $\mathcal{L}_2$, see below. The crucial point to observe is the fact that the transition

$$\mathcal{L}_{\text{QCD}} \Rightarrow \mathcal{L}_{\text{eff}}$$

is a non perturbative phenomenon. It is very different from the construction of the effective theory for the linear sigma model, where the loop expansion in the original theory does make sense, and where the low-energy representation can be worked out order by order in the loop expansion.

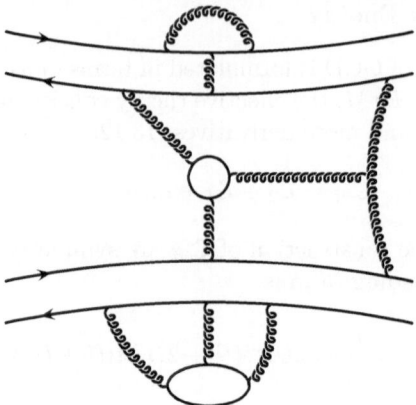

Fig. 7. $\pi\pi$ scattering in QCD. Displayed is one of the infinitely many graphs that contribute in QCD, and which are all equally important

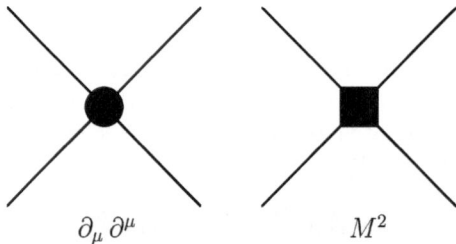

Fig. 8. Evaluating the $\pi\pi$ scattering amplitude with the effective lagrangian $\mathcal{L}_2$ in (19). Only the two graphs displayed contribute at leading order. The symbol M^2 is defined in (20)

I have provided

- no proof that $\mathcal{L}_2$ is the correct leading order lagrangian.
- no discussion of the group theoretical background (realization of the group $SU_R(2) \times SU_L(2)$ on curved manifolds).

For these issues, I refer the interested reader to earlier Schladming lectures by Leutwyler[14], Manohar[15] and Ecker[16]. Leutwyler has proven in [17] that the above lagrangian does reproduce the Green functions of QCD at low energy, see also below.

5   Calculations with $\mathcal{L}_{\text{eff}}$

In the last section, we have displayed the effective lagrangian that allows one to evaluate masses, form factors and scattering matrix elements in QCD at low energy. The results are valid at small values of the quark masses and of the momenta. In this section, I illustrate the procedure with several examples.

5.1   Leading Terms

Evaluating matrix elements at tree level with the effective lagrangian (19) generates the leading term in the low-energy expansion, see later for a precise meaning of this terminology.

Pion Masses. Consider the kinetic term in $\mathcal{L}_2$,

$$\mathcal{L}_2 = \frac{1}{2}[\partial_\mu \vec{\pi} \cdot \partial^\mu \vec{\pi} - M^2 \vec{\pi}^2] + O(\vec{\pi}^4) \ .$$

We conclude that the three pions have the same mass in this approximation,

$$M_{\pi^\pm}^2 = M_{\pi^0}^2 = M^2 \ . \tag{21}$$

Remark: We denote by $M_{\pi^\pm}$ and M_{π^0} the charged and neutral pion mass, respectively. The symbol M_π stands for the pion mass in the isospin symmetry limit, whereas M^2 is the first term in the quark mass expansion of the charged and neutral pion (mass)^2.

$\pi\pi$ **Scattering.** The terms of order $\vec{\pi}^4$ in $\mathcal{L}_2$ generate the leading term in the $\pi\pi$ scattering amplitude,

$$\mathcal{L}_2 = \text{kinetic term} + \frac{1}{8F^2}[\partial_\mu \vec{\pi}^2 \partial^\mu \vec{\pi}^2 - M^2 \vec{\pi}^4] + O(\vec{\pi}^6) \ .$$

The couplings are

$$\frac{1}{F^2} \quad \text{and} \quad \frac{M^2}{F^2} \ .$$

In other words, the mass M is also a coupling! Figure 8 displays the two diagrams that one has to evaluate in this case.

a) $\pi^+\pi^- \to \pi^0\pi^0$

The scattering matrix element is

$$T = \frac{M^2 - s}{F^2} + O(p^4) \ ; \quad s = (p_1 + p_2)^2 \ .$$

This result agrees with (1), up to the replacement $(M, F) \to (M_\pi, F_\pi)$, which is a higher order effect, included in the symbol $O(p^4)$.

b) $\pi^a \pi^b \to \pi^c \pi^d$

The structure of the matrix element in the general case is displayed in (12). The invariant amplitude $A(s,t,u)$ becomes

$$A(s,t,u) = \frac{s - M^2}{F^2} + O(p^4) \, . \tag{22}$$

c) Isospin amplitudes

For later use, we introduce here the isospin amplitudes

$$T^{I=0} = 3A(s,t,u) + A(t,u,s) + A(u,s,t) \, ,$$
$$T^{I=1} = A(t,u,s) - A(u,s,t) \, ,$$
$$T^{I=2} = A(t,u,s) + A(u,s,t) \, .$$

5.2 Higher Order Trees

Evaluating tree-level graphs with $\mathcal{L}_2$ generates the leading terms of the quantities in question. How about the contributions from $\mathcal{L}_4$? It contains e.g. terms with four derivatives, like

$$\mathcal{L}_4 = d_1 \langle \partial_\mu U \partial^\mu U^\dagger \rangle^2 + \cdots \, .$$

What is the effect of this term? From

$$\langle \partial_\mu U \partial^\mu U^\dagger \rangle^2 = \frac{4}{F^4} (\partial_\mu \vec{\pi} \cdot \partial^\mu \vec{\pi})^2 + \cdots \, ,$$

we conclude that $\mathcal{L}_4$ also contributes to elastic $\pi\pi$ scattering. Since four derivatives are involved, there will be four powers of the momenta. Indeed, at tree level, $\mathcal{L}_4$ contributes with

$$\delta A(s,t,u) = \frac{8d_1}{F^4}(s - 2M^2)^2 + \cdots \, . \tag{23}$$

This term is of order p^4 for small momenta. One can perform the expansion in a systematic manner - there are only a limited number of terms at order p^4 [12]:

$$\mathcal{L}_4 = \sum_{i=1}^{7} l_i P_i \, ,$$
$$P_1 = \frac{1}{4}\langle \partial_\mu U \partial^\mu U^\dagger \rangle^2 \, , \ P_2 = \frac{1}{4}\langle \partial_\mu U \partial_\nu U^\dagger \rangle \langle \partial^\mu U \partial^\nu U^\dagger \rangle \, ,$$
$$P_3 = \frac{1}{16}\langle 2B\mathcal{M}(U + U^\dagger)\rangle^2 \, , \ P_7 = -\frac{1}{16}\langle 2B\mathcal{M}(U - U^\dagger)\rangle^2 \, .$$

$P_{4,5,6}$ do not contribute to the $\pi\pi$ scattering amplitude. Why is knowledge of $\mathcal{L}_4$ useful? Suppose that the LECs l_i are known $\Rightarrow$ low-energy amplitudes are parametrized in terms of 7 parameters $\Rightarrow$ all other amplitudes fixed in terms of these parameters at this order in the low-energy expansion.

Comments

- Chiral symmetry and C, P, T invariance have been used to determine $\mathcal{L}_4$. For example,

$$\langle \partial_\mu U \partial^\mu U^\dagger \rangle^2$$

  is invariant under $U \to V_R U V_L^\dagger$, $V_I \in SU(2)$.
- The low-energy constants l_i (LECs) are not fixed by symmetry arguments.
- The operators $P_{1,2,3,7}$ contribute to the following processes:

$$
\begin{aligned}
P_1, P_2 &\to \pi\pi \to \pi\pi, \dots \\
P_3 &\to M_\pi^2, \pi\pi \to \pi\pi, \ \dots \\
P_7 &\to M_{\pi^+}^2 - M_{\pi^0}^2 \ .
\end{aligned}
$$

Example. Taking into account the contributions from $\mathcal{L}_4$, the pion masses read at tree level

$$
M_{\pi^\pm}^2 = M^2 + \frac{2l_3}{F^2} M^4 \ ,
$$

$$
M_{\pi^0}^2 = M_{\pi^\pm}^2 - \frac{2B^2}{F^2}(m_d - m_u)^2 l_7 \ . \tag{24}
$$

5.3 Loops

Scattering amplitudes, evaluated with

$$\mathcal{L}_2 + \mathcal{L}_4$$

at tree level, are still real. To be in accord with optical theorem, one needs to evaluate loops. This guarantees that unitarity is satisfied order by order in the low-energy expansion, like in any standard loop expansion in QFT. We illustrate the evaluation of loops in the case of the pion mass.

Pion Mass. To evaluate the pion mass in the isospin symmetry limit $m_u = m_d$, we consider the connected two-point function

$$
\delta^{ab} \triangle_c(p) = i \int d^4x \, e^{ipx} \langle 0|T\phi^a(x)\phi^b(0)|0\rangle_c \ . \tag{25}
$$

Fig. 9. Tree and tadpole contribution from $\mathcal{L}_2$ to the two-point function (25)

Taking into account the tree and tadpole contributions displayed in Fig. 9, we find that

$$\Delta_c(p) = \frac{Z}{M_p^2 - p^2} ,$$

where

$$M_p^2 = M^2 + \frac{M^2 I}{2F^2} , \quad I = \int \frac{d^4 l}{i(2\pi)^4} \frac{1}{M^2 - l^2} .$$

The tadpole integral I is quadratically divergent. A standard method to cope with this situation is to perform the integral in d dimensions,

$$I \to \int \frac{d^d l}{i(2\pi)^d} \frac{1}{M^2 - l^2} = \frac{M^{d-2}}{(4\pi)^{d/2}} \Gamma(1 - d/2) .$$

The result is now finite at $d \neq 2, 4, 6 \ldots$, whereas it is still still divergent at $d = 4$:

$$M^2 I \to \frac{M^4}{8\pi^2} \frac{1}{d - 4} , \quad d \to 4 .$$

There is also a tree contribution from $\mathcal{L}_4$, see (24),

$$\delta M_p^2 = \frac{2l_3}{F^2} M^4 .$$

If we tune l_3 to diverge as $d \to 4$ as well, we may cancel the divergence in the mass:

$$l_3 \to -\frac{1}{32\pi^2} \frac{1}{d - 4} + \text{finite} , \quad d \to 4$$

The physical pion mass at $m_u = m_d$ finally becomes

$$M_\pi^2 = M^2 - \frac{M^4}{32\pi^2 F^2} \bar{l}_3 + O(M^6) . \tag{26}$$

The quantity $\bar{l}_3$ contains the finite parts from l_3 and from the tadpole integral I [12].

Comments

- The required counterterm to cancel the divergence stems from $\mathcal{L}_4$. The divergences cannot be canceled by tuning parameters in $\mathcal{L}_2$: this is a typical feature of a non-renormalizable theory.
- Non-renormalizability does not mean non-calculability: the effective theory of QCD allows one to calculate all quantities perfectly well. The only disadvantage is the increasing number of low-energy coupling constants that one is faced with while incorporating higher order terms in the expansion.
- The contribution from the tadpole is suppressed, it is of order M^4, not of order M^2.

5.4   Renormalization Made Systematic

Similar divergences occur in other amplitudes, like $\pi\pi \to \pi\pi$, when loops are calculated. These divergences can all be canceled by tuning the l_i in the lagrangian with four derivatives. The final prescription for the evaluation of the Green functions at order p^4 reads as follows: the effective lagrangian is

$$\mathcal{L}_{\text{eff}} = \mathcal{L}_2 + \mathcal{L}_4 + \mathcal{L}_6 + \cdots ,$$

where $\mathcal{L}_2$ is given in (19), and

$$\mathcal{L}_4 = \sum_{i=1}^{7} l_i P_i \, , \ \mathcal{L}_6 = \sum_{i=1}^{53} c_i Q_i \, .$$

The polynomials $P_i (Q_i)$ and the divergent parts in the LECs $l_i (c_i)$ have been determined in [12] ([18]). At order p^6, there are in addition so called parity odd terms, that come with an epsilon tensor [19]. Further, some of the polynomials P_i, Q_i vanish in the absence of external fields. I refer the reader to [18] for a guided tour through $\mathcal{L}_6$.

The Green functions are calculated as follows:

leading terms: trees with $\mathcal{L}_2$

next-to-leading terms: $\begin{cases} \text{trees with} & \mathcal{L}_4 \\ \text{one loop with} & \mathcal{L}_2 \end{cases}$

next-to-next-to-leading terms: $\begin{cases} \text{trees with} & \mathcal{L}_6 \\ \text{one loop with} & \mathcal{L}_4 \\ \text{two loops with} & \mathcal{L}_2 \end{cases}$

Comments

- Next-to-leading terms are suppressed by p^2 with respect to the leading term, the next-to-next-to leading terms by p^4, etc.
- The finite parts of the low-energy constants l_i, c_i are not fixed by symmetry considerations. On the other hand, they are calculable in principle in QCD. We are witnessing a transition period, where one starts to be able to evaluate the LECs in the framework of QCD - see [20] for recent examples.
- Meanwhile, until the transition period is over, one takes LECs from data, from sum rules etc.
- We will comment in the final section about the introduction of additional fundamental fields in the effective lagrangian.

6   Pion-Pion Scattering

In this section, I consider in some detail the evaluation of the elastic $\pi\pi$ scattering amplitude in the low-energy region at order p^6. The motivation for investigating this reaction in great detail includes the following points:

- It is possible to make precise theoretical predictions.
- Some of the threshold amplitudes are sensitive to the mechanism of spontaneous chiral symmetry breaking.
- There are new experimental activities:

  BNL-865 at Brookhaven K_{e4} decays [21]
  NA48 at CERN K_{e4} decays [22]
  KLOE at DAFNE K_{e4} decays [23]
  DIRAC at CERN $\pi^+\pi^-$ atom [24]
- The analysis of the P-wave amplitude in elastic $\pi\pi$ scattering provides very useful information for the evaluation of the anomalous magnetic moment of the muon [25].

Why do the so called K_{e4} decays

$$K^+ \to \pi^+\pi^- e^+\nu_e \,,$$
$$K^+ \to \pi^0\pi^0 e^+\nu_e \,,$$
$$K^0 \to \pi^0\pi^- e^+\nu_e \,,$$

and their charge conjugate modes provide information on $\pi\pi$ scattering?

First Explanation. The emerging pions in the decay products interact with each other (*final state interaction*). The decay width is therefore sensitive to this interaction.

Second Explanation (More Learned). The decay matrix element is described by form factors. One may perform a partial wave analysis of these - the partial wave amplitudes then carry the phases of the $\pi\pi$ interaction (Watson's final state theorem). In the decay, one measures interference terms between the various partial waves. The decay of the charged kaon into a charged pion pair plus leptons turns out to be sensitive to

$$\delta_{l=0}^{I=0}(E_{\pi\pi}) - \delta_1^1(E_{\pi\pi}) \,,$$

where $E_{\pi\pi}$ denotes the centre-of-mass energy of the pion pair. The investigation of $\pi^+\pi^-$ atoms (Pionium) is of interest for the following reason. The atom is formed by electromagnetic interactions. It is not stable - the ground state, e.g., has various decay channels,

$$A_{\pi^+\pi^-} \to \pi^0\pi^0, \gamma\gamma, \pi^0\pi^0\gamma\gamma, \dots \,.$$

The decay into the neutral pion pair depends on the strength of the strong interactions [26,27],

$$\Gamma = \frac{2}{9}\alpha^3 p^* |a_0^0 - a_0^2|^2 (1+\delta) \,, \tag{27}$$

where a_0^0 denotes the $I = 0$ S-wave scattering length $a_{l=0}^{I=0}$, and p^* is the modulus of the centre-of-mass momentum of the neutral pions in the rest system of the decaying atom. Further, $\alpha \simeq 1/137$ is the fine structure constant of QED. The decay width into a photon pair is suppressed by the factor $\alpha^{3/2}$ with respect to the $\pi^0\pi^0$ channel. The correction δ is also known [27],

$$\delta = 0.058 \pm 0.012 \,.$$

A measurement of the lifetime of the ground state therefore amounts to a measurement of the combination $|a_0^0 - a_0^2|$.

6.1   Chiral Expansion of the $\pi\pi$ Amplitude

The chiral expansion of the invariant scattering amplitude $A(s,t,u)$ is performed in the manner discussed in the previous section and has the structure

$$A(s,t,u) = A_2 + A_4 + A_6 + \cdots \,,$$

where A_{2n} is of order p^{2n}.

Leading Order. The leading order result is given in (22). For the $I = 0$ S-wave scattering length, one obtains from that expression

$$a_0^0 = \frac{7M^2}{32\pi F^2} = 0.16 \,. \tag{28}$$

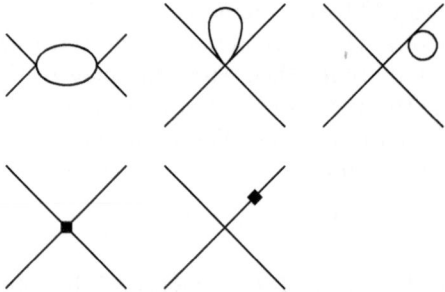

Fig. 10. One-loop graphs in elastic $\pi\pi$ scattering. The *filled squares* denote contributions from $\mathcal{L}_4$

In the numerical evaluation, I have replaced M and F by $M_{\pi\pm} = 139.57$ MeV and by $F_\pi = 92.4$ MeV, respectively (the replacements amount to taking into account some of the higher order effects in the calculation of the scattering amplitude). The data on K_{e4} decays collected in the seventies gave [28]

$$a_0^0 = 0.26 \pm 0.05 \qquad [\text{from 30'000 } K_{e4} \text{ decays}] .$$

The question was – for many years – whether this indicates a failure of the chiral prediction (28). Indeed, if the condensate $\langle 0|\bar{u}u|0\rangle$ is small or vanishing, one can understand a large value of the scattering length. This is the main idea of the so-called *generalized chiral perturbation theory*, which was much discussed at the end of the last millennium [29]. In order to decide the issue, one needs more precise data and a more precise calculation. Both is available in our days, as we will show below.

Next-to-Leading Order. One evaluates one-loop graphs with $\mathcal{L}_2$ and tree-graphs with $\mathcal{L}_4$. Some of these are displayed in Fig. 10. The result can be written in the form

$$A_4 = a_1 M_\pi^4 + a_2 M_\pi^2 s + a_3 s^2 + a_4 (t-u)^2$$
$$+ F(s) + G(s,t) + G(s,u) .$$

We note the following:

- The amplitude A_4 carries four powers of momenta
- F, G denote loop functions. They generate the imaginary parts which are required by unitarity at order p^4. Example:

$$F(s) = \frac{1}{2F_\pi^4}(s^2 - M_\pi^4)\bar{J}(s) ,$$

$$\bar{J}(s) = \frac{1}{16\pi^2}\left\{\sigma \log \frac{1-\sigma}{1+\sigma} + 2 + i\pi\sigma\right\} ,$$

$$\sigma = (1 - 4M_\pi^2/s)^{1/2} \ ; \ s > 4M_\pi^2 .$$

The function $\bar{J}(s)$ is analytic in the complex s-plane, cut along the real axis for $s \geq 4M_\pi^2$.

- a_i contain the low-energy constants $\bar{l}_{1,2,3,4}$ from $\mathcal{L}_4$. For example, the $I = 0$ S-wave scattering length now reads

$$a_0^0 = \frac{7}{32\pi} \frac{M_\pi^2}{F_\pi^2} \left\{ 1 + \varepsilon + O(M_\pi^4) \right\} ,$$

$$\varepsilon = \frac{5}{84\pi^2} \frac{M_\pi^2}{F_\pi^2} \left(\bar{l}_1 + 2\bar{l}_2 - \frac{3}{8}\bar{l}_3 + \frac{21}{10}\bar{l}_4 + \frac{21}{8} \right) .$$

- the LECs $\bar{l}_i$ contain so-called *chiral logarithms*:

$$\bar{l}_i \to -\log M_\pi^2 , \quad M_\pi \to 0 ,$$

which generate large corrections,

$$a_0^0 = \frac{7M_\pi^2}{32\pi F_\pi^2} \left\{ 1 - \underbrace{\frac{9}{32\pi^2} \frac{M_\pi^2}{F_\pi^2} \log M_\pi^2/\mu^2}_{\chi\text{logarithm}} + \cdots \right\} .$$

The χ logarithms generate a 25% correction at $\mu = 1$ GeV.

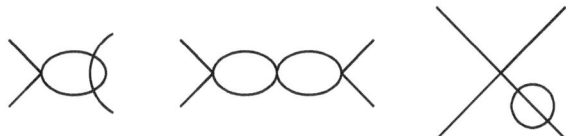

Fig. 11. Two-loop graphs in elastic $\pi\pi$ scattering

Next-to-Next-to Leading Order. In order to evaluate A_6, one needs to perform a two-loop calculation with $\mathcal{L}_2$, one-loop with $\mathcal{L}_4$ and trees with $\mathcal{L}_6$. A few of the two-loop graphs are displayed in Fig. 11. It turns out that all graphs can be evaluated in closed form [30]! The structure of A_6 is as follows:

$A_6 = $ loop functions
$$+b_1 M_\pi^6 + b_2 M_\pi^4 s + b_3 M_\pi^2 s^2 + b_4 M_\pi^2 (t-u)^2 + b_5 s^3 + b_6 s(t-u)^2 .$$

The calculation determines $b_1, \ldots, b_6$ in terms of the LECs at order p^4 and p^6 [30]. Obviously, one is faced with a problem here: one needs to determine the LECs before a precise prediction of the scattering lengths can be performed.

6.2   Type A,B LECs

One encounters the following LECs in the $\pi\pi$ amplitude up to and including terms of order p^6:

$$\bar{l}_1, \bar{l}_2, \bar{l}_3, \bar{l}_4 : \quad \mathcal{L}_4$$
$$\bar{r}_1, ..., \bar{r}_6 \quad : \quad \mathcal{L}_6$$

They come in two categories [31]:

1. *Terms that survive in the chiral limit:*

$$\bar{l}_1, \bar{l}_2, \bar{r}_5, \bar{r}_6 \quad \text{type A}$$

These LECs show up in momentum dependence of the $\pi\pi$ amplitude, and may therefore be determined phenomenologically.

2. *Symmetry breaking terms:*

$$\bar{l}_3, \bar{l}_4, \bar{r}_1, \bar{r}_2, \bar{r}_3, \bar{r}_4 \quad \text{type B}$$

These LECs specify the quark mass dependence of the amplitude. The $\pi\pi$ scattering amplitude cannot provide information on type B LECs, because one cannot vary the quark mass in experiments.

We discuss in the following section how these LECs can be determined, and what the result for the scattering lengths finally is.

7   Roy Equations and Threshold Parameters

Roy equations allow one to determine type A LECs and to pin down the $\pi\pi$ scattering amplitude in the low-energy region with high precision. It is useful to discuss Roy equations in two steps. In the first step, no relation to chiral symmetry is used - the only ingredients are analyticity, unitarity, crossing symmetry, data above 800 MeV and Regge behaviour beyond 2 GeV. In a second step, one merges the representation of the amplitude with the chiral representation discussed in the previous section.

7.1 $\pi\pi \to \pi\pi$: Roy I

Analyticity, crossing symmetry and the Froissart bound lead to dispersion relations for the partial waves of the $\pi\pi$ scattering amplitude. As has been shown by Roy [32], the imaginary part of the partial waves is needed only in the physical region. These dispersion relations read

$$t_\ell^I(s) = k_\ell^I(s) + \sum_{I'=0}^{2} \sum_{\ell'=0}^{\infty} \int_{4M_\pi^2}^{\infty} ds' \, K_{\ell\ell'}^{II'}(s, s') \, \mathrm{Im}\, t_{\ell'}^{I'}(s') \, ,$$

where $\ell(I)$ denotes the angular momentum (isospin) of the partial wave. The subtraction term $k_\ell^I(s)$ contains the scattering lengths a_0^0, a_0^2, and the kernels

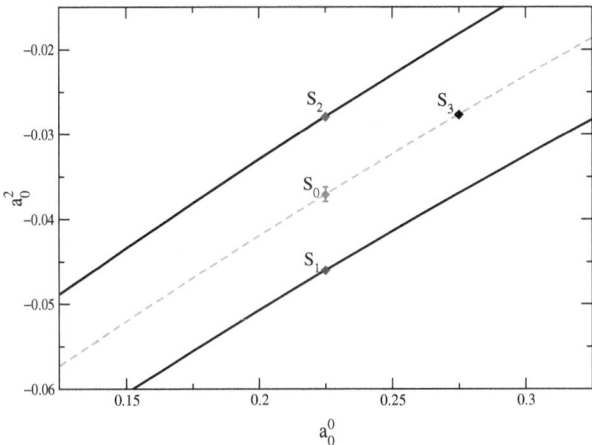

Fig. 12. The universal band in the $a_0^0 - a_0^2$ plane. For each point in this band, a unique solution to the Roy equations can be constructed

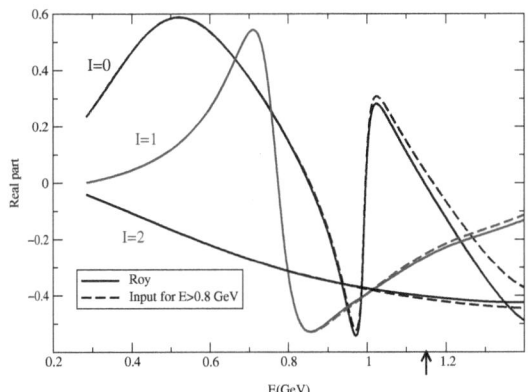

Fig. 13. Example of a solution to the Roy equations. Displayed are the real parts of the partial waves, as a function of the centre-of-mass energy of the pions

$K_{\ell\ell'}^{II'}(s, s')$ are explicitly known functions of s and of s'. At low energy, only S- and P- waves matter, and the contributions from the remaining waves may be expanded in a Taylor series. Unitarity expresses the imaginary parts of the partial waves in terms of the real parts (in the elastic region) – the above dispersion relations then become (singular) integral equations for the two S- and for the P-wave. These may be solved numerically [33,35]. In [35], it has been shown that the two S-wave scattering lengths a_0^0, a_0^2 are the essential parameters. Once these are fixed, experimental data (above 800 MeV) plus Regge behaviour above 2 GeV determine the amplitude in the low energy region to within rather small uncertainties. Figure 12 displays the *universal band* in the $a_0^0 - a_0^2$ plane, for which solutions may be found. An example solution for the three waves is displayed in Fig. 13.

7.2 $\pi\pi \to \pi\pi$: Roy II

In the previous subsection, no reference to ChPT was made. We now invoke this information: one requires that the χ amplitude agrees with phenomenological amplitude below the threshold region. Subtractions and matching are performed in such a manner that χ logarithms are suppressed. This procedure allows one to determine the type A LECs listed in Sect. 6.2. Type B LECs are determined form other sources:

$$\bar{l}_4 :\leftrightarrow \quad \text{scalar radius of pions}$$
$$\bar{l}_3 :\leftrightarrow \quad \text{SU(3), Zweig rule}$$
$$\bar{r}_{1,2,3,4} \leftrightarrow \text{resonance saturation}$$

We have now arrived at solutions to the Roy equations that agree with the chiral amplitude at low energy. In other words, the three lowest partial waves below 800 MeV are fixed. The scattering lengths of the partial waves with $\ell \geq 1$, as well as the effective ranges (also those of the S-waves) can be expressed in terms of sum rules over the imaginary parts [36]. In Fig. 14 is displayed the resulting P-wave phase shift. This allows one to pin down the electromagnetic form factor of the pion with high precision [25]. This form factor plays a very important role in the evaluation of the anomalous magnetic moment of the muon, see Marc Knecht's lectures at this school.

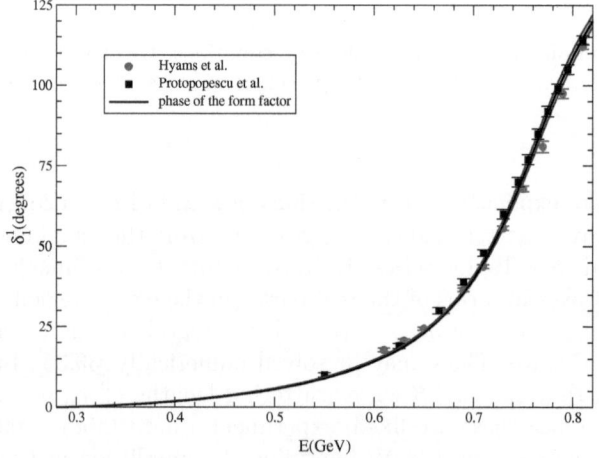

Fig. 14. The P-wave from the Roy analysis

Finally, we come to the two S-wave scattering lengths, which are now also fixed [37,31],

$$a_0^0 = 0.220 \pm 0.005 ,$$
$$a_0^0 - a_0^2 = 0.265 \pm 0.004 .$$

I refer the reader to [31] for scattering lengths and effective ranges of other waves. The above result also provides [27] a precise prediction for the lifetime of Pionium in the ground state via (27),

$$\tau = (2.9 \pm 0.1) \times 10^{-15} \text{ sec} .$$

This prediction is presently confronted with experiment at DIRAC [24].

7.3   The Coupling $\bar{l}_3$

The main difference between generalized chiral perturbation theory (GChPT [29]) and the standard picture used here resides in the coupling constant $\bar{l}_3$, which can take any value in GChPT. Let me recall where this constant occurs. First, consider the chiral expansion of the pion mass (26). We write

$$\bar{l}_3 = \log \frac{\Lambda_3^2}{M_\pi^2} .$$

Crude estimates in the standard version of ChPT give [12]

$$0.2 \text{ GeV} < \Lambda_3 < 2 \text{ GeV} .$$

The term of order M^4 in (26) is then very small compared to the leading term, i.e., the Gell-Mann-Oakes-Renner formula is obeyed very well. As mentioned, GChPT allows for arbitrarily large values of $\bar{l}_3$. The quadratic term in (26) is then not leading, the series must be reordered. It is very satisfactory that experiment can decide the issue, for the following reason. The constant $\bar{l}_3$ also occurs in the expression for the scattering lengths a_0^0 and a_0^2. One may then perform the matching of the chiral and the phenomenological amplitude as discussed above with $\bar{l}_3$ as a free parameter. The result of this investigation [31,38] is displayed in Fig. 15 [39]. The allowed values of the scattering lengths lie in the small hatched band in the figure. This band reflects a low-energy theorem [31] for the difference $2a_0^0 - 5a_0^2$,

$$2a_0^0 - 5a_0^2 = \frac{3M_\pi^2}{4\pi F_\pi^2}\left(1 + \frac{M_\pi^2 \langle r^2 \rangle_s}{3} + \frac{41M_\pi^2}{192\pi^2 F_\pi^2} + O(M_\pi^4)\right) , \qquad (29)$$

where $\langle r^2 \rangle_s$ denotes the scalar radius of the pion. The terms of order M_π^6 in (29) generate the curvature of the band. For each value in the band, one can calculate the difference $\delta_0^0 - \delta_1^1$ of $\pi\pi$ phase shifts, and compare with

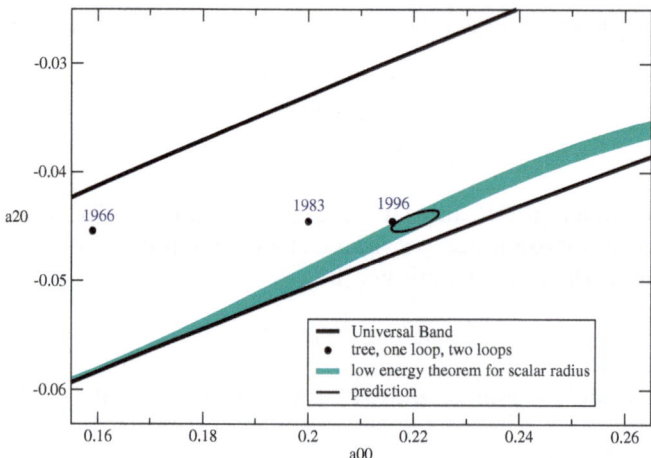

Fig. 15. Constraints imposed on the S-wave scattering lengths by chiral symmetry. The three circles illustrate the prediction of the chiral perturbation theory at increasing order. The error ellipse represents the final result from [31], while the narrow, curved band indicates the region allowed in GChPT

data on $K^+ \to \pi^+\pi^- e^+\nu_e$ decays. E865 has performed this analysis, with the result [21]

$$a_0^0 = 0.216 \pm 0.013\,(\text{stat}) \pm 0.002\,(\text{syst}) \pm 0.002\,(\text{theor}) .$$

The central value leads to $\bar{l}_3 \simeq 6$, with an uncertainty of about 10 units. In the expansion (26), the second term amounts to a correction of about four percent at $\bar{l}_3 = 6$. From this, one concludes that the first term in the mass expansion dominates by far: the motivation for a generalized scheme [29], with a small quark condensate and a large second term in (26) has evaporated, at least for $SU(2) \times SU(2)$.

7.4   Note Added: On the Precision of the Theoretical Predictions for $\pi\pi$ Scattering

In two recent papers [40], Peláez and Ynduráin evaluate some of the low energy observables of $\pi\pi$ scattering and obtain flat disagreement with our results [31] that I have described above. The authors work with unsubtracted dispersion relations, so that their results are very sensitive to the poorly known high energy behaviour of the scattering amplitude. They claim that the asymptotic representation we used in [35,31] is incorrect and propose an alternative one. We have repeated [41] their calculations on the basis of the standard, subtracted fixed-t dispersion relations, using their asymptotics. The outcome fully confirms our earlier findings. Moreover, we show that the Regge parametrization proposed by these authors for the region above 1.4

GeV violates crossing symmetry: Their ansatz is not consistent with the behaviour observed at low energies.

8 Outlook

Instead of a summary, I have provided at the school an outlook on topics not covered in the lectures. This outlook was based on the structure of the effective chiral lagrangian of the Standard Model [42] displayed in Table 1. The numbers in brackets denote the number of independent LECs[2]. The numbers refer to $N_f = 3$, except for the pieces with superscript πN.

Table 1. The effective lagrangian of the Standard Model [42]

$$\mathcal{L}_{\text{chiral order}} \quad (\text{\# of LECs})$$

$$\mathcal{L}_{p^2}(2) + \mathcal{L}_{p^4}^{\text{odd}}(0) + \mathcal{L}_{G_F p^2}^{\Delta S=1}(2) + \mathcal{L}_{e^2 p^0}^{\text{em}}(1) + \mathcal{L}_{G_8 e^2 p^0}^{\text{emweak}}(1)$$

$$+ \mathcal{L}_p^{\pi N}(1) + \mathcal{L}_{p^2}^{\pi N}(7) + \mathcal{L}_{G_8 p^0}^{MB, \Delta S=1}(2) + \mathcal{L}_{G_8 p}^{MB, \Delta S=1}(8) + \mathcal{L}_{e^2 p^0}^{\pi N, \text{em}}(3)$$

$$+ \underline{\mathcal{L}_{p^4}^{\text{even}}(10)} + \underline{\mathcal{L}_{p^6}^{\text{odd}}(32)} + \underline{\mathcal{L}_{G_8 p^4}^{\Delta S=1}(22)} + \underline{\mathcal{L}_{e^2 p^2}^{\text{em}}(14)} + \underline{\mathcal{L}_{G_8 e^2 p^2}^{\text{emweak}}(14)}$$

$$+ \underline{\mathcal{L}_{e^2 p}^{\text{leptons}}(5)}$$

$$+ \underline{\mathcal{L}_{p^3}^{\pi N}(23)} + \underline{\mathcal{L}_{p^4}^{\pi N}(118)} + \mathcal{L}_{G_8 p^2}^{MB, \Delta S=1}(?) + \underline{\mathcal{L}_{e^2 p}^{\pi N, \text{em}}(8)}$$

$$+ \underline{\mathcal{L}_{p^6}^{\text{even}}(90)}$$

The underlined lagrangians are fully renormalized: their divergence structure has been determined in a process independent manner. In the lectures, I had discussed aspects of $\mathcal{L}_{p^2}$, $\mathcal{L}_{p^4}^{\text{even}}$ and of $\mathcal{L}_{p^6}^{\text{even}}$ in the $SU(2) \times SU(2)$ case. The above table opens a very wide field of further applications:

Topics. Meson and baryon decays: electromagnetic, semileptonic, leptonic, non leptonic, rare and not so rare; decay constants F_π, F_K, ...; scattering amplitudes: $\pi\pi \to \pi\pi, \pi N \to \pi N, \gamma N \to \pi N, ...$; mixing angles; quark mass ratios; large N_C investigations; anomalies; isospin violation; weak matrix elements; quark condensate; hadronic atoms; lattice: $m_q \to 0, V \to \infty$; quenched ChPT.

---

[2] In $\mathcal{L}_{p^4}^{\pi N}$, I changed the number 114 in Eckers table into 118, in order to agree with [43]. I thank Ulf-G. Meißner for pointing out the correct number of LECs in this case.

In order to study nuclear physics in the framework of ChPT, the above lagrangian must still be enlarged. A vast amount of massive calculations in this topic have been performed by V. Bernard, E. Epelbaum, W. Glöckle, N. Kaiser, Ulf-G. Meißner and others, see e.g. [44]. This method has allowed one to put theoretical nuclear physics on a sound basis.

For recent contributions to the topics just mentioned, see [45,46].

Acknowledgements

It is a pleasure to thank the organizers of this Winter School for the warm hospitality, for the perfect organization of the School and for the very pleasant weather conditions. In addition, I thank Christoph Häfeli and Martin Schmid for providing me with many figures that I displayed during the lectures, and which are now incorporated partly also here. Finally, I thank Ulf-G. Meißner for carefully reading the manuscript. This work was supported in part by the Swiss National Science Foundation and by RTN, BBW-Contract No. 01.0357 and EC-Contract HPRN–CT2002–00311 (EURIDICE).

References

1. S. Weinberg, Phys. Rev. Lett. **17**, 616 (1966).
2. S. Scherer, in *Advances in Nuclear Physics, Vol. 27*, edited by J. W. Negele and E. W. Vogt (Kluwer Academic/Plenum Publishers, New York, 2003) [arXiv:hep-ph/0210398].
3. K. Hagiwara *et al.* [Particle Data Group Collaboration], Phys. Rev. D **66**, 010001 (2002).
4. Y. Nambu, Phys. Rev. Lett. **4**, 380 (1960); Phys. Rev. **117**, 648 (1960);
   Y. Nambu and G. Jona-Lasinio, Phys. Rev. **122**, 345 (1961); ibid. **124**, 246 (1961).
5. J. Goldstone, Nuovo Cim. **19**, 154 (1961);
   The paper by J. Goldstone, A. Salam and S. Weinberg, Phys. Rev. **127**, 965 (1962), gives three different proofs of the theorem.
6. H. Leutwyler, Phys. Lett. B **378**, 313 (1996) [arXiv:hep-ph/9602366].
7. C. Vafa and E. Witten, Nucl. Phys. B **234**, 173 (1984).
8. H. Araki and R. Haag, Comm. Math. Phys. **4**, 77 (1967).
9. H. Lehmann, K. Symanzik and W. Zimmermann, Nuovo Cim. **1**, 205 (1955).
10. W. E. Caswell and G. P. Lepage, Phys. Lett. B **167**, 437 (1986).
11. A. Nyffeler and A. Schenk, Annals Phys. **241**, 301 (1995) [arXiv:hep-ph/9409436].
12. J. Gasser and H. Leutwyler, Annals Phys. **158**, 142 (1984).
13. S. Weinberg, Physica A **96**, 327 (1979).
14. H. Leutwyler, *Chiral Effective Lagrangians*, in: Recent Aspects of Quantum Fields, Proceedings of the XXX. Internationale Universitätswochen für Kern- und Teilchenphysik, Schladming, Austria, Feb. 27 - March 8, 1991, Springer Lecture Notes in Physics, Vol. 396, H. Mitter, H. Gausterer (eds.).

15. A. V. Manohar, Effective Field Theories, Lect. Notes Phys. **479** [arXiv:hep-ph/9606222].

16. G. Ecker, Chiral Symmetry, Lect. Notes Phys **521** [arXiv:hep-ph/9805500].

17. H. Leutwyler, Annals Phys. **235**, 165 (1994) [arXiv:hep-ph/9311274].

18. J. Bijnens, G. Colangelo and G. Ecker, JHEP **9902**, 020 (1999) [arXiv:hep-ph/9902437]; Annals Phys. **280**, 100 (2000) [arXiv:hep-ph/9907333].

19. T. Ebertshauser, H. W. Fearing and S. Scherer, Phys. Rev. D **65**, 054033 (2002) [arXiv:hep-ph/0110261];
J. Bijnens, L. Girlanda and P. Talavera, Eur. Phys. J. C **23**, 539 (2002) [arXiv:hep-ph/0110400].

20. J. Heitger, R. Sommer and H. Wittig [ALPHA Collaboration], Nucl. Phys. B **588**, 377 (2000) [arXiv:hep-lat/0006026];
A. C. Irving, C. McNeile, C. Michael, K. J. Sharkey and H. Wittig [UKQCD Collaboration], Phys. Lett. B **518**, 243 (2001) [arXiv:hep-lat/0107023];
D. R. Nelson, G. T. Fleming and G. W. Kilcup, Nucl. Phys. Proc. Suppl. **106**, 221 (2002) [arXiv:hep-lat/0110112];
G. T. Fleming, D. R. Nelson and G. W. Kilcup, Nucl. Phys. Proc. Suppl. **119** (2003) 245 [arXiv:hep-lat/0209141];
F. Farchioni, C. Gebert, I. Montvay and L. Scorzato [qq+q Collaboration], arXiv:hep-lat/0209142;
F. Farchioni, C. Gebert, I. Montvay, E. Scholz and L. Scorzato [qq+q Collaboration], Phys. Lett. B **561**, 102 (2003) [arXiv:hep-lat/0302011];
F. Farchioni, I. Montvay, E. Scholz and L. Scorzato [qq+q Collaboration], Eur. Phys. J. C **31** (2003) 227 [arXiv:hep-lat/0307002].

21. S. Pislak *et al.* [BNL-E865 Collaboration], Phys. Rev. Lett. **87**, 221801 (2001) [arXiv:hep-ex/0106071]; Phys. Rev. D **67**, 072004 (2003) [arXiv:hep-ex/0301040].

22. R. Batley *et al.* [NA48-Collaboration], Addendum III (to Proposal P253/ CERN/SPSC) for a Precision Measurement of Charged Kaon Decay Parameters with an extended NA48 Setup, CERN/SPSC/P253 add. 3, Jan. 16, 2000.

23. L. Maiani, G. Pancheri and N. Paver (eds.), *The Second* DAΦNE *Physics Handbook* (INFN-LNF-Divisione Ricerca, SIS-Ufficio Publicazioni, Frascati, 1995).

24. B. Adeva *et al.*, CERN proposal CERN/SPSLC 95-1, 1995.

25. H. Leutwyler, *Electromagnetic form factor of the pion*, in: Continuous Advances in QCD 2002: Arkadyfest – honoring the 60th birthday of Prof. Arkady Vainshtein, K. A. Olive, M. A. Shifman and M. B. Voloshin (eds.), World Scientific, 2003, [arXiv:hep-ph/0212324].

26. S. Deser, M. L. Goldberger, K. Baumann and W. Thirring, Phys. Rev. **96**, 774 (1954).

27. J. Gasser, V. E. Lyubovitskij, A. Rusetsky and A. Gall, Phys. Rev. D **64**, 016008 (2001) [arXiv:hep-ph/0103157];
J. Gasser, V. E. Lyubovitskij and A. Rusetsky, Phys. Lett. B **471**, 244 (1999) [arXiv:hep-ph/9910438];
H. Sazdjian, Phys. Lett. B **490**, 203 (2000) [arXiv:hep-ph/0004226]; arXiv:hep-ph/0012228.
Recent work on related matters is described in
A. Gashi, G. Rasche, G. C. Oades and W. S. Woolcock, Nucl. Phys. A **628**, 101 (1998) [arXiv:nucl-th/9704017];

H. Jallouli and H. Sazdjian, Phys. Rev. D **58**, 014011 (1998) [arXiv:hep-ph/9706450];

P. Labelle and K. Buckley [arXiv:hep-ph/9804201];

M.A. Ivanov, V.E. Lyubovitskij, E.Z. Lipartia and A.G. Rusetsky, Phys. Rev. D **58**, 094024 (1998) [arXiv:hep-ph/9805356];

P. Minkowski, in: Proceedings of the International Workshop Hadronic Atoms and Positronium in the Standard Model, Dubna, 26-31 May 1998, M.A. Ivanov *at al.* (eds.), Dubna 1998 [arXiv:hep-ph/9808387];

E.A. Kuraev, Phys. Atom. Nucl. **61**, 239 (1998);

U. Jentschura, G. Soff, V. Ivanov and S.G. Karshenboim, Phys. Lett. A **241**, 351 (1998);

B.R. Holstein, Phys. Rev. D **60**, 114030 (1999) [arXiv:nucl-th/9901041];

X. Kong and F. Ravndal, Phys. Rev. D **59**, 014031 (1999); ibid. D **61**, 077506 (2000) [arXiv:hep-ph/9905539];

H. W. Hammer and J. N. Ng, Eur. Phys. J. A **6**, 115 (1999) [arXiv:hep-ph/9902284];

D. Eiras and J. Soto, Phys. Rev. D **61**, 114027 (2000) [arXiv:hep-ph/9905543]; Phys. Lett. B **491**, 101 (2000) [arXiv:hep-ph/0005066].

28. L. Rosselet *et al.*, Phys. Rev. D **15**, 574 (1977).

29. M. Knecht, B. Moussallam, J. Stern and N. H. Fuchs, Nucl. Phys. B **457**, 513 (1995) [arXiv:hep-ph/9507319]; ibid. B **471**, 445 (1996) [arXiv:hep-ph/9512404].

30. J. Bijnens, G. Colangelo, G. Ecker, J. Gasser and M. E. Sainio, Phys. Lett. B **374**, 210 (1996) [arXiv:hep-ph/9511397]; Nucl. Phys. B **508**, 263 (1997) [Erratum-ibid. B **517**, 639 (1998)] [arXiv:hep-ph/9707291].

31. G. Colangelo, J. Gasser and H. Leutwyler, Nucl. Phys. B **603**, 125 (2001) [arXiv:hep-ph/0103088].

32. S. M. Roy, Phys. Lett. B **36**, 353 (1971).

33. J. L. Basdevant, J. C. Le Guillou and H. Navelet, Nuovo Cim. A **7**, 363 (1972); M.R. Pennington and S.D. Protopopescu, Phys. Rev. D **7**, 1429 (1973); ibid. D **7**, 2591 (1973);

J. L. Basdevant, C. D. Froggatt and J. L. Petersen, Phys. Lett. B **41**, 173 (1972); ibid. 178; ibid. B **72**, 413 (1974);

J. L. Petersen, Acta Phys. Austriaca Suppl. **13**, 291 (1974); Yellow report CERN 77-04 (1977);

C. D. Froggatt and J. L. Petersen, Nucl. Phys. B **91**, 454 (1975); ibid. B **104**, 186 (1976) (E); ibid. B **129**, 89 (1977);

D. Morgan and M.R. Pennington, in [23], p. 193;

P. Büttiker, *Comparison of Chiral Perturbation Theory with a Dispersive Analysis of $\pi\pi$ Scattering*, PhD thesis, Universität Bern, 1996;

B. Ananthanarayan and P. Büttiker, Phys. Rev. D **54**, 1125 (1996) [arXiv:hep-ph/9601285]; Phys. Rev. D **54**, 5501 (1996) [arXiv:hep-ph/9604217]; Phys. Lett. B **415**, 402 (1997) [arXiv:hep-ph/9707305] and in [34], p. 370; O. O. Patarakin, V. N. Tikhonov and K. N. Mukhin, Nucl. Phys. A **598**, 335 (1996);

O. O. Patarakin (for the CHAOS collaboration), in [34], p. 376, and arXiv:hep-ph/9711361;

M. Kermani *et al.* [CHAOS Collaboration], Phys. Rev. C **58**, 3431 (1998);

B. Loiseau, R. Kaminski and L. Lesniak, πN Newslett. **16**, 349 (2002) [arXiv:hep-ph/0110055];

R. Kaminski, L. Lesniak and B. Loiseau, arXiv:hep-ph/0207063; Phys. Lett. B **551**, 241 (2003) [arXiv:hep-ph/0210334].

34. A.M. Bernstein, D. Drechsel and T. Walcher (eds.), *Chiral Dynamics: Theory and Experiment*, Workshop held in Mainz, Germany, 1-5 Sept. 1997, Lecture Notes in Physics Vol. 513, Springer, 1997.

35. B. Ananthanarayan, G. Colangelo, J. Gasser and H. Leutwyler, Phys. Rept. **353**, 207 (2001) [arXiv:hep-ph/0005297].

36. G.Wanders, Helv. Phys. Acta **39**, 228 (1966).

37. G. Colangelo, J. Gasser and H. Leutwyler, Phys. Lett. B **488**, 261 (2000) [arXiv:hep-ph/0007112].

38. G. Colangelo, J. Gasser and H. Leutwyler, Phys. Rev. Lett. **86**, 5008 (2001) [arXiv:hep-ph/0103063].

39. I thank H. Leutwyler for providing me with this figure.

40. J. R. Peláez and F. J. Ynduráin, Phys. Rev. D **68** (2003) 074005 [arXiv:hep-ph/0304067]; F. J. Ynduráin, arXiv:hep-ph/0310206.

41. I. Caprini, G. Colangelo, J. Gasser and H. Leutwyler, Phys. Rev. D **68** (2003) 074006 [arXiv:hep-ph/0306122].

42. G. Ecker, *Strong interactions of light flavours*, in: Advanced School on QCD, Benasque, Spain, July 2000, S. Peris and V. Vento (eds.), Univ. Autonoma des Barcelona, Servei de Publicacions, Bellaterra (Barcelona), 2001 [arXiv:hep-ph/0011026]. The table was downloaded with permission from G. Ecker.

43. N. Fettes, U. G. Meißner, M. Mojžiš and S. Steininger, Annals Phys. **283**, 273 (2000); Erratum-ibid. **288**, 249 (2001) [arXiv:hep-ph/0001308].

44. U. G. Meißner, V. Bernard, E. Epelbaum and W. Glöckle, *Recent results in chiral nuclear dynamics*, Invited talk at 2002 International Workshop on Strong Coupling Gauge Theories and Effective Field Theories (SCGT 02), Nagoya, Japan, 10-13 Dec 2002, arXiv:nucl-th/0301079.

45. Proceedings of the 3rd Workshop on Chiral Dynamics - Chiral Dynamics 2000: Theory and Experiment, Newport News, Virginia, 17-22 Jul 2000; Published by World Scientific, 2002, (Proceedings from the Institute for nuclear Theory, vol. 11), A. M. Bernstein, J. L. Goity and U. -G. Meißner (eds.) [arXiv:hep-ph/0011140].

46. J. Bijnens, U. G. Meißner and A. Wirzba, *Effective Field Theories of QCD*, to appear in: Proceedings of 264th WE-Heraeus Seminar: Workshop on Effective Field Theories of QCD, Bad Honnef, Germany, 26-30 Nov 2001 [arXiv:hep-ph/0201266].

The Anomalous Magnetic Moment
of the Muon: A Theoretical Introduction

Marc Knecht

Centre de Physique Théorique, CNRS Luminy, Case 907, 13288 Marseille
Cedex 9, France, knecht@cpt.univ-mrs.fr

1 Introduction

In February 2001, the Muon (g-2) Collaboration of the E821 experiment at
the Brookhaven AGS released a new value of the anomalous magnetic moment
of the muon a_μ, measured with an unprecedented accuracy of 1.3 ppm [parts
per million]. This annoucement has caused quite some excitement in the
particle physics community. Indeed, this experimental value was claimed to
show a deviation of 2.6 σ with one of the most accurate evaluations of the
anomalous magnetic moment of the muon within the standard model. It was
subsequently shown that a sign error in one of the theoretical contributions
was responsible for a sizeable part of this discrepancy, which eventually only
amounted to 1.6 σ. However, this event had the merit to draw the attention
to the fact that low energy but high precision experiments represent real
potentialities, complementary to the high energy accelerator programs, for
evidencing possible new degrees of freedom, supersymmetry or whatever else,
beyond those described by the standard model of electromagnetic, weak, and
strong interactions.

Clearly, in order for theory to match such an accurate measurement, calculations in the standard model have to be pushed to their very limits. The
difficulty is not only one of having to compute higher orders in perturbation
theory, but also to correctly take into account strong interaction contributions involving low-energy scales, where non perturbative effects are important,
and which therefore represent a real theoretical challenge. Furthermore, in
the meantime the members of experimental team at Brookhaven have further
improved upon their measurement. In July 2002, they have announced a new
result, in perfect agreement with the one they had obtained about one year
and a half before, but with an error lowered down to the 0.7 ppm level. In
addition, the theoretical evaluation of a specific and important contribution,
called hadronic vacuum polarization, on which we shall have more to say later on, has shown a discrepancy between different data sets that are used as
imputs. Depending on the choice between these conflicting data, the discrepancy between the experimental and the theoretical values can be as small
as 1 σ, certainly not a case for beyond the standard model physics, or can

M. Knecht, The Anomalous Magnetic Moment of the Muon: A Theoretical Introduction, Lect.
Notes Phys. **629**, 37–84 (2004)
http://www.springerlink.com/   © Springer-Verlag Berlin Heidelberg 2004

reach the 3 σ level, a much more promising situation as far as the possibility of "new physics" is concerned.

The purpose of this account is to give an overview of the main features of the theoretical calculations that have been done in order to obtain accurate predictions for the anomalous magnetic moment of the muon within the standard model. Actually, all three charged leptons, $e^{\pm}$, $\mu^{\pm}$, and $\tau^{\pm}$, of the standard model can be treated on the same footing, except that the very different values of their masses will induce different sensitivities with respect to the mass scales involved in the higher order quantum corrections. Thus, the anomalous magnetic moment of the electron is almost only [but not quite] sensitive to the electromagnetic interactions of the leptons, and its value is barely affected by strong interaction effects or by weak interaction corrections. On the other hand, the strengths of the latter two types of corrections are enhanced by a considerable factor, $\sim (m_\tau/m_e)^2 \sim 1.2 \times 10^7$, in the case of the τ, as compared to the electron. The same huge enhancement factor would also affect the contributions coming from degrees of freedom beyond the standard model, so that the measurement of the anomalous magnetic moment of the τ would represent the best opportunity to detect new physics. Unfortunately, the very short lifetime of the τ lepton which, precisely because of its high mass, can also decay into hadronic states, makes such a measurement impossible at present. The muon lies somewhat in the intermediate range of mass scales [1] and its lifetime still makes a measurement of its anomalous magnetic moment possible. However, in order to obtain an accurate [that is, comparable to the experimental accuracy] prediction, the contributions of all the sectors of the standard model have to be known very precisely. Therefore, although the other two lepton flavours will also be discussed, since this does not really require additional work, the emphasis of these lectures will nevertheless be put on the muon.

There exist several excellent reviews and introductions, which the interested reader may consult. As far as the situation up to 1990 is concerned, the collection of articles published in [1] offers a wealth of information, on both theory and experiment. Very useful accounts of earlier theoretical work are presented in [2,3]. Among the more recent reviews, [4–8] are most informative. I shall not touch the subject of the study of new physics scenarios which might offer an explanation for a possible deviation between the standard model prediction of the magnetic moment of the muon and its experimental value, should such a deviation be confirmed in the future. For this aspect, I refer the reader to [9] and to the articles quoted therein, or to [10] for a list of the recent papers on the subject.

---

[1] The corresponding enhancement factor is $\sim (m_\mu/m_e)^2 \sim 4 \times 10^4$.

2   General Considerations

In the context of relativistic quantum mechanics, the interaction of a pointlike spin one-half particle of charge e_ℓ and mass m_ℓ [ℓ stands hereafter for any of the three charged lepton flavours e, μ or τ] with an external electromagnetic field $\mathcal{A}_\mu(x)$ is described by the Dirac equation with the minimal coupling prescription,

$$i\hbar \frac{\partial \psi}{\partial t} = \left[c\boldsymbol{\alpha} \cdot \left(-i\hbar\boldsymbol{\nabla} - \frac{e_\ell}{c}\boldsymbol{\mathcal{A}} \right) + \beta m_\ell c^2 + e_\ell \mathcal{A}_0 \right] \psi \,. \tag{2.1}$$

In the non relativistic limit, this reduces to the Pauli equation for the two-component spinor φ describing the large components of the Dirac spinor ψ,

$$i\hbar \frac{\partial \varphi}{\partial t} = \left[\frac{(-i\hbar\boldsymbol{\nabla} - (e_\ell/c)\boldsymbol{\mathcal{A}})^2}{2m_\ell} - \frac{e_\ell\hbar}{2m_\ell c}\boldsymbol{\sigma} \cdot \mathbf{B} + e_\ell \mathcal{A}_0 \right] \varphi \,. \tag{2.2}$$

As is well known, this equation amounts to associate with the particle's spin a magnetic moment

$$\mathbf{M}_s = g_\ell \left(\frac{e_\ell}{2m_\ell c} \right) \mathbf{S} \,,\ \mathbf{S} = \hbar \frac{\boldsymbol{\sigma}}{2} \,, \tag{2.3}$$

with a gyromagnetic ratio predicted to be $g_\ell = 2$.

In the context of quantum field theory, the response to an external electromagnetic field is described by the matrix element of the electromagnetic current [2] $\mathcal{J}^\rho$ [spin projections and Dirac indices of the spinors are not written explicitly]

$$\langle \ell^-(p')|\mathcal{J}^\rho(0)|\ell^-(p)\rangle = \bar{u}(p')\Gamma^\rho(p',p)u(p) \,, \tag{2.4}$$

with [$k_\mu \equiv p'_\mu - p_\mu$]

$$\Gamma^\rho(p',p) = F_1(k^2)\gamma^\rho + \frac{i}{2m_\ell}F_2(k^2)\sigma^{\rho\nu}k_\nu - F_3(k^2)\gamma_5\sigma^{\rho\nu}k_\nu$$
$$+ F_4(k^2)[k^2\gamma^\rho - 2m_\ell k^\rho]\gamma_5 \,. \tag{2.5}$$

This expression of the matrix element $\langle \ell^-(p')|\mathcal{J}^\rho(0)|\ell^-(p)\rangle$ is the most general that follows from Lorentz invariance, the Dirac equation for the two spinors, $(\not{p} - m)u(p) = 0$, $\bar{u}(p')(\not{p}' - m) = 0$, and the conservation of the electromagnetic current, $(\partial \cdot \mathcal{J})(x) = 0$. The two first form factors, $F_1(k^2)$ and $F_2(k^2)$, are known as the Dirac form factor and the Pauli form factor, respectively. Since the electric charge operator $\mathcal{Q}$ is given, in units of the charge e_ℓ, by

$$\mathcal{Q} = \int d\mathbf{x}\, \mathcal{J}_0(x^0, \mathbf{x}) \,, \tag{2.6}$$

---

[2] In the standard model, $\mathcal{J}^\rho$ denotes the total electromagnetic current, with the contributions of all the charged elementary fields in presence, leptons, quarks, electroweak gauge bosons,...

the form factor $F_1(k^2)$ is normalized by the condition $F_1(0) = 1$. The presence of the form factor $F_3(k^2)$ requires both parity and time reversal invariance to be broken, whereas $F_4(k^2)$ can be different from zero provided parity is broken. Both $F_3(k^2)$ and $F_4(k^2)$ are therefore absent if only electromagnetic and strong interactions are considered [we leave aside the possibility of having a non vanishing vacuum angle in the strong interaction sector]. On the other hand, in the standard model, the weak interactions violate both parity and time reversal symmetry, so that they actually induce non vanishing expressions for these form factors.

The above form factors are defined for $k^2 < 0$, and they are real in this region if the current $\mathcal{J}_\rho(x)$ is hermitian. Due to general properties of quantum field theory, like causality, analyticity, and crossing symmetry, these form factors can be analytically continued into the whole complex k^2 plane with a cut for $k^2 > 4m_\ell^2$. They then become complex functions, obeying the Schwartz reflection property $F_i(k^2)^* = F_i(k^{2*})$. For $k^2 > 4m_\ell^2$, the form factors $F_i(k^2 + i\epsilon)$ describe the crossed channel matrix element $\langle \ell^-(p')\ell^+(p)|\mathcal{J}^\rho(0)|0\rangle$. Furthermore, at $k^2 = 0$, they describe the residue of the s-channel photon pole in the S-matrix element for elastic $\ell^+\ell^-$ scattering.

At tree level in the standard model, one finds

$$F_1^{\text{tree}}(k^2) = 1, \ F_i^{\text{tree}}(k^2) = 0, \ i = 2, 3, 4. \tag{2.7}$$

In order to obtain non zero values for $F_2(k^2)$, $F_3(k^2)$, and $F_4(k^2)$ already at tree level, the interaction of the Dirac field with the photon field $\mathcal{A}_\mu$ would have to depart from the minimal coupling prescription. For instance, the modification $[\mathcal{F}_{\mu\nu} = \partial_\mu \mathcal{A}_\nu - \partial_\nu \mathcal{A}_\mu, \ \mathcal{J}^\rho = \overline{\psi}\gamma^\rho\psi]$

$$\int d^4x \mathcal{L}_{\text{int}} = -\frac{e_\ell}{c} \int d^4x \mathcal{J}^\rho \mathcal{A}_\rho \ \rightarrow$$

$$\int d^4x \widehat{\mathcal{L}}_{\text{int}} = -\frac{e_\ell}{c} \int d^4x \left[\mathcal{J}^\rho \mathcal{A}_\rho + \frac{\hbar}{4m_\ell} a_\ell \overline{\psi}\sigma_{\mu\nu}\psi \mathcal{F}^{\mu\nu} + \frac{\hbar}{2e_\ell} d_\ell \overline{\psi} i\gamma_5 \sigma_{\mu\nu}\psi \mathcal{F}^{\mu\nu} \right]$$

$$= -\frac{e_\ell}{c} \int d^4x \widehat{\mathcal{J}}^\rho \mathcal{A}_\rho, \tag{2.8}$$

with [3]

$$\widehat{\mathcal{J}}_\rho = \mathcal{J}_\rho - \frac{\hbar}{2m_\ell} a_\ell \partial^\mu \left(\overline{\psi}\sigma_{\mu\rho}\psi \right) - \frac{\hbar d_\ell}{e_\ell} \partial^\mu \left(\overline{\psi} i\gamma_5 \sigma_{\mu\rho}\psi \right), \tag{2.9}$$

leads to

$$\widehat{F}_1^{\text{tree}}(k^2) = 1, \ \widehat{F}_2^{\text{tree}}(k^2) = a_\ell, \ \widehat{F}_3^{\text{tree}}(k^2) = d_\ell/e_\ell, \ \widehat{F}_4^{\text{tree}}(k^2) = 0. \tag{2.10}$$

---

[3] The current $\widehat{\mathcal{J}}^\rho$ is still a conserved four-vector, therefore the matrix element $\langle \ell^-(p')|\widehat{\mathcal{J}}^\rho(0)|\ell^-(p)\rangle$ also takes the form (2.4), (2.5), with appropriate form factors $\widehat{F}_i(k^2)$.

The equation satisfied by the Dirac spinor ψ then reads

$$i\hbar \frac{\partial \psi}{\partial t} = \left[c\boldsymbol{\alpha} \cdot \left(-i\hbar \boldsymbol{\nabla} - \frac{e_\ell}{c} \boldsymbol{\mathcal{A}} \right) + \beta m_\ell c^2 + e_\ell \mathcal{A}_0 \right.$$
$$\left. + \frac{e_\ell \hbar}{2m_\ell} a_\ell \beta \left(i\boldsymbol{\alpha} \cdot \mathbf{E} - \boldsymbol{\Sigma} \cdot \mathbf{B} \right) - \hbar d_\ell \beta \left(\boldsymbol{\Sigma} \cdot \mathbf{E} + i\boldsymbol{\alpha} \cdot \mathbf{B} \right) \right] \psi, \quad (2.11)$$

and the corresponding non relativistic limit becomes [4]

$$i\hbar \frac{\partial \varphi}{\partial t} = \left[\frac{(-i\hbar \boldsymbol{\nabla} - (e_\ell/c)\boldsymbol{\mathcal{A}})^2}{2m_\ell} - \frac{e_\ell \hbar}{2m_\ell c} (1 + a_\ell) \boldsymbol{\sigma} \cdot \mathbf{B} - \hbar d_\ell \boldsymbol{\sigma} \cdot \mathbf{E} \right.$$
$$\left. + e_\ell \mathcal{A}_0 + \cdots \right] \varphi. \quad (2.12)$$

Thus the coupling constant a_ℓ induces a shift in the gyromagnetic factor, $g_\ell = 2(1 + a_\ell)$, while d_ℓ gives rise to an electric dipole moment. The modification (2.8) of the interaction with the photon field introduces two arbitrary constants, and both terms produce a *non renormalizable* interaction. Non constant values of the form factors could be generated at tree level upon introducing [11] additional non renormalizable couplings, involving derivatives of the external field of the type $\Box^n \mathcal{A}_\mu$, which preserve the gauge invariance of the corresponding field equation satisfied by ψ. In a similar way, one can also introduce terms which induce a nonzero value for F_4. In a renormalizable framework, like QED or the standard model, calculable non vanishing values for $F_2(k^2)$, $F_3(k^2)$, and $F_4(k^2)$ are generated by the loop corrections. In particular, the latter will likewise induce an *anomalous magnetic moment*

$$a_\ell = \frac{1}{2}(g_\ell - 2) = F_2(0) \quad (2.13)$$

and an electric dipole moment $d_\ell = e_\ell F_3(0)$; $F_4(0)$, which corresponds to an axial radius of the lepton, is also called the *anapole moment* [12–14], and is sensitive to the gradients of the external fields.

If we consider only the electromagnetic and the strong interactions, the current $\mathcal{J}^\rho$ is gauge invariant, and the two form factors that remain in that case, symmetry $F_1(k^2)$ and $F_2(k^2)$, do not depend on the gauges chosen in order to quantize the photon and the gluon gauge fields. This is no longer the case if the weak interactions are included as well, since $\mathcal{J}^\rho$ now transforms in a non trivial way under a weak gauge transformation, and the corresponding form factors in general depend on the gauge choices. As we have already mentioned above, the zero momentum transfer values $F_i(0)$, $i = 1, 2, 3, 4$ describe a physical S-matrix element. To the extent that the perturbative S-matrix

---

[4] Terms involving the gradients of the external fields $\mathbf{E}$ and $\mathbf{B}$ or terms nonlinear in these fields are not shown.

of the standard model does not depend on the gauge fixing parameters to any order of the renormalized perturbation expansion, the quantities $F_i(0)$ should define *bona fide* gauge-fixing independent observables.

The computation of $\Gamma_\rho(p',p)$ is often a tedious task, especially if higher loop contributions are considered. It is therefore useful to concentrate the efforts on computing the form factor of interest, e.g. $F_2(k^2)$ in the case of the anomalous magnetic moment. This can be achieved upon projecting out the different form factors [15,16] using the following general expression [5]

$$F_i(k^2) = \text{tr}\left[\Lambda_i^\rho(p',p)(\not{p}' + m_\ell)\Gamma_\rho(p',p)(\not{p} + m_\ell)\right],\qquad (2.14)$$

with

$$\Lambda_1^\rho(p',p) = \frac{1}{4}\frac{1}{k^2 - 4m_\ell^2}\gamma^\rho + \frac{3m_\ell}{2}\frac{1}{(k^2 - 4m_\ell^2)^2}(p' + p)^\rho$$

$$\Lambda_2^\rho(p',p) = -\frac{m_\ell^2}{k^2}\frac{1}{k^2 - 4m_\ell^2}\gamma^\rho - \frac{m_\ell}{k^2}\frac{k^2 + 2m_\ell^2}{(k^2 - 4m_\ell^2)^2}(p' + p)^\rho$$

$$\Lambda_3^\rho(p',p) = -\frac{i}{2k^2}\frac{1}{k^2 - 4m_\ell^2}\gamma_5(p' + p)^\rho$$

$$\Lambda_4^\rho(p',p) = -\frac{1}{4k^2}\frac{1}{k^2 - 4m_\ell^2}\gamma_5\gamma^\rho.\qquad (2.15)$$

For $k \to 0$, one has

$$\Lambda_2^\rho(p',p) = \frac{1}{4k^2}\left[\gamma^\rho - \frac{1}{m_\ell}\left(1 + \frac{k^2}{m_\ell^2}\right)(p + \frac{1}{2}k)^\rho + \cdots\right],\qquad (2.16)$$

and

$$(\not{p} + m_\ell)\Lambda_2^\rho(p',p)(\not{p}' + m_\ell) = \frac{1}{4}(\not{p} + m_\ell)\left[-\frac{k^\rho}{k^2}\right.$$

$$\left. + (\gamma^\rho - \frac{p^\rho}{m_\ell})\frac{\not{k}}{k^2} + \cdots\right].\qquad (2.17)$$

The last expression behaves as $\sim 1/k$ as the external photon four momentum k_μ vanishes, so that one may worry about the finiteness of $F_2(0)$ obtained upon using (2.14). This problem is solved by the fact that $\Gamma^\rho(p',p)$ satisfies the Ward identity

$$(p' - p)_\rho\Gamma^\rho(p',p) = 0,\qquad (2.18)$$

---

[5] From now on, I most of the time use the system of units where $\hbar = 1$, $c = 1$. Other projectors on $F_2(k^2)$ have also been devised, see e.g. [17], but are not currently used.

following from the conservation of the electromagnetic current. Therefore, the identity

$$\Gamma^\rho(p',p) = -k_\sigma \frac{\partial}{\partial k_\rho} \Gamma^\sigma(p',p) \qquad (2.19)$$

provides the additional power of k which ensures a finite result as $k_\mu \to 0$.

The presence of three different interactions in the standard model naturally leads one to consider the following decomposition of the anomalous magnetic moment a_ℓ:

$$a_\ell = a_\ell^{\text{QED}} + a_\ell^{\text{had}} + a_\ell^{\text{weak}} . \qquad (2.20)$$

The first term, a_ℓ^{QED}, denotes all the contributions which arise from loops involving only virtual photons and leptons. Among these, it is useful to distinguish those which involve only the same lepton flavour ℓ for which we wish to compute the anomalous magnetic moment, and those which involve loops with leptons of different flavours, denoted collectively as ℓ' [$\alpha \equiv e^2/4\pi$],

$$a_\ell^{\text{QED}} = \sum_{n \geq 1} A_n \left(\frac{\alpha}{\pi}\right)^n + \sum_{n \geq 2} B_n(\ell, \ell') \left(\frac{\alpha}{\pi}\right)^n . \qquad (2.21)$$

The second type of contribution, a_ℓ^{had}, involves also quark loops. Their contribution is far from being limited to the short distance scales, and a_ℓ^{had} is an intrinsically non perturbative quantity. From a theoretical point of view, this represents a serious difficulty. Finally, at some level of precision, the weak interactions can no longer be ignored, and contributions of virtual Higgs or massive gauge boson degrees of freedom induce the third component a_ℓ^{weak}. Of course, starting from the two loop level, a hadronic contribution to a_ℓ^{weak} will also be present. The remainder of this presentation is devoted to a detailed discussion of these various contributions.

Before starting this guided tour of the anomalous magnetic moments of the massive charged leptons of the standard model, it is useful to keep in mind a few simple considerations:

• The anomalous magnetic moment is a dimensionless quantity. Therefore, the coefficients A_n above are *universal*, i.e. they do not depend on the flavour of the lepton whose anomalous magnetic moment we wish to evaluate.

• The contributions to a_ℓ of degrees of freedom corresponding to a typical scale $M \gg m_\ell$ decouple [18], i.e. they are *suppressed* by powers of m_ℓ/M.[6]

---

[6] In the presence of the weak interactions, this statement has to be reconsidered. Indeed, the necessity for the cancellation of the $SU(2) \times U(1)$ gauge anomalies [19–21] transforms the decoupling of, say, a single heavy fermion in a given generation, into a somewhat subtle issue [22,23], the resulting lagrangian being no longer renormalizable.

• The contributions to a_ℓ originating from light degrees of freedom, characterized by a typical scale $m \ll m_\ell$ are *enhanced* by powers of $\ln(m_\ell/m)$. At a given order, the logarithmic terms that do not vanish as $m_\ell/m \to 0$ can often be computed from the knowledge of the lower order terms and of the β function through the renormalization group equations [24–27].

These general properties already allow to draw several elementary conclusions. The electron being the lightest charged lepton, its anomalous magnetic moment is dominantly determined by the values of the coefficients A_n. The first contribution of other degrees of freedom comes from graphs involving, say, at least one muon loop, which occurs first at the two-loop level, and is of the order of $(m_e/m_\mu)^2(\alpha/\pi)^2 \sim 10^{-10}$. The hadronic effects, i.e. "quark and gluon loops", characterized by a scale of ~ 1 GeV, or effects of degrees of freedom beyond the standard model, which may appear at some high scale M, will be felt more strongly, by a considerable factor $(m_\mu/m_e)^2 \sim 40\,000$, in a_μ than in a_e. Thus, a_e is well suited for testing the validity of QED at higher orders, whereas a_μ is more appropriate for testing the weak sector of the standard model, one of the main motivations for the BNL experiment, and possibly for detecting new physics.

Exercises for Section 2

Exercise 2.1
Show that the expression of the matrix element of the electromagnetic current given by (2.4) and (2.5) indeed follows from the conditions stated. Show that the form factors $F_i(k^2)$ in these equations are real if the current $\mathcal{J}^\rho$ is hermitian. Work out the transformation properties of the different form factors under the operations of parity and time reversal. How many additional form factors are needed in order to describe the same matrix element if the assumption concerning the conservation of the current is dropped?

Exercise 2.2
Show that the current $\widehat{\mathcal{J}}^\rho$ defined by (2.8) is conserved.

Exercise 2.3
Find the term one needs to add to $\widehat{\mathcal{L}}_{\text{int}}$, and thus to $\widehat{\mathcal{J}}_\rho$, such as to generate a constant but nonzero form factor $\widehat{F}_4$ at tree level. Show that in the non relativistic limit it induces an interaction term of the form $F_4(0)\boldsymbol{\sigma} \cdot (\boldsymbol{\nabla} \wedge \mathbf{B})$ in a non uniform magnetic field.

Exercise 2.4
Show that the quantities $\Lambda_i^\rho(p',p)$ defined in (2.15) indeed project on the corresponding form factors $F_i(k^2)$ through (2.14). Derive (2.16) and (2.17).

Exercise 2.5
Work out the expression of the electromagnetic current in the standard model.

Exercise 2.6
Give the most general decomposition of the matrix element in (2.5) in the case of a massive Majorana neutrino [hint: see [28,29] for a rather complete treatment].

3 Brief Overview of the Experimental Situation

3.1 Measurements of the Magnetic Moment of the Electron

The first indication that the gyromagnetic factor of the electron is different from the value $g_e = 2$ predicted by the Dirac theory came from the precision measurement of hyperfine splitting in hydrogen and deuterium [30]. The first measurement of the gyromagnetic factor of free electrons was performed in 1958 [31], with a precision of 3.6%. The situation began to improve with the introduction of experimental setups based on the Penning trap. Some of the successive values obtained over a period of forty years are shown in Table 1. Technical improvements, eventually allowing for the trapping of a single electron or positron, produced, in the course of time, an enormous increase in precision which, starting from a few percents, went through the ppm levels, before culminating at 4 ppb [parts per billion] in the last [32] of a series of experiments performed at the University of Washington in Seattle. The same experiment has also produced a measurement of the magnetic moment of the positron with the same accuracy, thus providing a test of CPT invariance at the level of 10^{-12},

$$g_{e^-}/g_{e^+} = 1 + (0.5 \pm 2.1) \times 10^{-12}. \tag{3.1}$$

Assuming invariance under CPT, the weighted average of the electron and positron anomalous moments obtained in [32] gives [33,34],

$$a_e^{\mathrm{exp}} = 0.001\,159\,652\,188\,3(4\,2). \tag{3.2}$$

An extensive survey of the literature and a detailed description of the various experimental aspects can be found in [35]. The earlier experiments are reviewed in [36].

3.2 Measurements of the Magnetic Moment of the Muon

The anomalous magnetic moment of the muon has also been the subject of quite a few experiments. The very short lifetime of the muon, $\tau_\mu = (2.19703 \pm 0.00004) \times 10^{-6}s$, makes it necessary to proceed in a completely

Table 1. Some experimental determinations of the electron's anomalous magnetic moment a_e with the corresponding relative precision.

0.001 19(5)	4.2%	[37]
0.001 165(11)	1%	[38]
0.001 116(40)	3.6%	[31]
0.001 160 9(2 4)	2 100 ppm	[39]
0.001 159 622(27)	23 ppm	[40]
0.001 159 660(300)	258 ppm	[41]
0.001 159 657 7(3 5)	3 ppm	[42]
0.001 159 652 41(20)	172 ppb	[43]
0.001 159 652 188 4(4 3)	4 ppb	[32]

different way in order to attain a high precision. The experiments conducted at CERN during the years 1968-1977 used a muon storage ring [for details, see [44] and references quoted therein]. The more recent experiments at the AGS in Brookhaven are based on the same concept. Pions are produced by sending a proton beam on a target. The pions subsequently decay into longitudinally polarized muons, which are captured inside a storage ring, where they follow a circular orbit in the presence of both a uniform magnetic field and a quadrupole electric field, the latter serving the purpose of stabilizing the orbits. The difference between the spin precession frequency and the orbit frequency is given by

$$\boldsymbol{\omega}_s - \boldsymbol{\omega}_c = -\frac{e}{m_\mu c}\left\{a_\mu \mathbf{B} - \left[a_\mu + \frac{1}{1-\gamma^2}\right]\boldsymbol{\beta} \wedge \mathbf{E}\right\},\qquad(3.3)$$

where $\boldsymbol{\beta}$ is the velocity of the muons, and γ is the corresponding Lorentz boost factor. Therefore, if γ is tuned to its "magic" value $\gamma = \sqrt{1 + 1/a_\mu} = 29.3$, the measurement of $\omega_s - \omega_c$ and of the magnetic field $\mathbf{B}$ allows to determine a_μ. The spin direction of the muon is determined by detecting the electrons or positrons produced in the decay of the muons with an energy greater than some threshold energy E_t. The number of detected electrons $N_e(t)$ decreases exponentially with time, the time constant being set by the muon's lifetime $\gamma\tau_\mu c$ in the laboratory frame, and is modulated by the frequency $\omega_s - \omega_c$,

$$N_e(t) = N_0(E_t)e^{-t/\gamma\tau_\mu}\{1 + A(E_t)\cos[(\omega_s - \omega_c)t + \phi(E_t)]\}.\qquad(3.4)$$

The observation of this time dependence thus provides the required measurement of $\omega_s - \omega_c$.

Several experimental results for the anomalous magnetic moment of the positively charged muon, obtained at the CERN PS or, more recently, at the BNL AGS, are recorded in Table 2. Notice that the relative errors are measured in ppm units, to be contrasted with the ppb level of accuracy achieved in the electron case. The four last values in Table 2 were obtained by the E821

Table 2. Determinations of the anomalous magnetic moment of the positively charged muon from the storage ring experiments conducted at the CERN PS and at the BNL AGS.

0.001 166 16(31)	265 ppm	[45]
0.001 165 895(27)	23 ppm	[46]
0.001 165 911(11)	10 ppm	[47]
0.001 165 925(15)	13 ppm	[48]
0.001 165 919 1(5 9)	5 ppm	[49]
0.001 165 920 2(1 6)	1.3 ppm	[50]
0.001 165 920 3(8)	0.7 ppm	[51]

experiment at BNL. They show a remarkable stability and a steady increase in precision, and now completely dominate the world average value. Further data, for negatively charged muons [7] are presently being analysed. The aim of the Brookhaven Muon (g - 2) Collaboration is to reach a precision of 0.35 ppm, but this will depend on whether the experiment will receive financial support to collect more data or not [8].

For completeness, one should mention that (3.3) is only correct as long as the muon has no electric dipole moment. If this is not the case, the more general relation,

$$\boldsymbol{\omega}_s - \boldsymbol{\omega}_c = -\frac{e}{m_\mu c}\left\{a_\mu \mathbf{B} - \left[a_\mu + \frac{1}{1-\gamma^2}\right]\boldsymbol{\beta}\wedge\mathbf{E}\right\} - \frac{2d_\mu}{\hbar}\left\{\boldsymbol{\beta}\wedge\mathbf{B}+\mathbf{E}\right\},$$

which holds for $\boldsymbol{\beta}\cdot\mathbf{E} = \boldsymbol{\beta}\cdot\mathbf{B} = 0$, has to be used. The additional term proportional to d_μ induces an oscillation of the muon spin with respect to the plane of motion [9]. In the standard model, given the experimental precision and the intensities of the fields used in the experiment, it is quite legitimate to use the formula (3.3). However, in case a discrepancy arises between the experimental value of a_μ and the standard model prediction, the difference could be induced by non standard contributions to either the anomalous

---

[7] The CERN experiment had also measured $a_{\mu^-} = 0.001\,165\,937(12)$ with a 10 ppm accuracy, giving the average value $a_\mu = 0.001\,165\,924(8.5)$, with an accuracy of 7 ppm.

[8] At the time of writing, the prospects in this respect are unfortunately rather dim.

[9] A measurement of the electric dipole moment of the muon was actually performed by the CERN experiment [47], with the result $d_\mu = (3.7\pm3.4)\times10^{-19}$ ecm. An even smaller value, $d_\mu \lesssim 9.1\times10^{-25}$ ecm, can be infered from the experimental value of the electric dipole moment of the electron [52,53], $d_e = 1.8(1.2)(1.0)\times10^{-27}$ ecm, and assuming a scaling law $d_\mu \sim \frac{m_\mu}{m_e}d_e$. Such a scaling law holds within the standard model, but not in models with flavour violating interactions, see for instance [54]. For a proposal to measure d_μ at the level of $\sim 10^{-24}$ ecm, see [55].

magnetic moment a_μ or the electric dipole moment d_μ, see the discussion in [54].

3.3   Experimental Bounds on the Anomalous Magnetic Moment of the τ Lepton

As already mentioned, the very short lifetime of the τ precludes a measurement of its anomalous magnetic moment following any of the techniques described above. Indirect access to a_τ is provided by the reaction $e^+e^- \rightarrow \tau^+\tau^-\gamma$. The results obtained by OPAL [56] and L3 [57] at LEP only lead to very loose bounds,

$$-0.052 < a_\tau < 0.058 \ (95\%C.L.)$$
$$-0.068 < a_\tau < 0.065 \ (95\%C.L.)\,, \tag{3.5}$$

respectively.

We shall now turn towards theory, in order to see how the standard model predictions compare with these experimental values. Only the cases of the electron and of the muon will be treated in some detail. The theoretical aspects as far as the anomalous magnetic moment of the τ are concerned are discussed in [58] and in the references quoted therein.

4   The Anomalous Magnetic Moment of the Electron

We start with the anomalous magnetic moment of the lightest charged lepton, the electron. Since the electron mass m_e is much smaller than any other mass scale present in the standard model, the mass independent part of a_e^{QED} dominates its value. As mentioned before, non vanishing contributions appear at the level of the loop diagrams shown in Fig. 1.

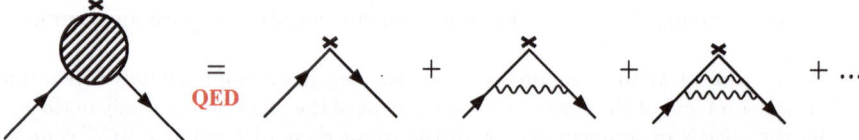

Fig. 1. The perturbative expansion of $\Gamma^\rho(p',p)$ in single flavour QED. The tree graph gives $F_1 = 1$, $F_2 = 0$, whereas $F_3(k^2)$ and $F_4(k^2)$ vanish identically to all orders in pure QED. The one loop vertex correction graph gives the coefficient A_1 in (2.21). The cross denotes the insertion of the external field.

4.1 The Lowest Order Contribution

The one loop diagram gives

$$\Gamma^\rho(p',p)\big|_{1 \text{ loop}} = (-ie)^2 \int \frac{d^4q}{(2\pi)^4} \gamma^\mu \frac{i}{p\!\!\!/' + q\!\!\!/ - m_e} \gamma^\rho \frac{i}{p\!\!\!/ + q\!\!\!/ - m_e} \gamma^\nu$$
$$\times \frac{(-i)}{q^2} \left[\eta_{\mu\nu} - (1 - \xi) \frac{q_\mu q_\nu}{q^2} \right]. \tag{4.1}$$

The photon propagator has been written in a Lorentz type gauge, corresponding to a covariant gauge fixing term $-(\partial \cdot A)^2/2\xi$. Let us for a moment
concentrate on the ξ - dependence of this one loop expression. Recall that
$\Gamma^\rho(p',p)$ has eventually to be inserted between the spinors $\bar{u}(p')$ and $u(p)$.
Then, the gauge dependence of the integrand is given by

$$(1 - \xi)\bar{u}(p') \, q\!\!\!/ \frac{i}{p\!\!\!/' + q\!\!\!/ - m_e} \gamma^\rho \frac{i}{p\!\!\!/ + q\!\!\!/ - m_e} \, q\!\!\!/ u(p) =$$
$$(1 - \xi)\bar{u}(p') i\gamma_\rho i u(p) \,,$$

and thus affects $F_1(k^2)$, but not $F_2(k^2)$. For evaluating the latter, one may
therefore take, say, $\xi = 1$ for convenience. The form factor $F_2(k^2)$ is then
obtained by using (2.14) and (2.15) and, upon evaluating the corresponding
trace of Dirac matrices, one finds

$$F_2(k^2)\big|_{1 \text{ loop}} =$$
$$ie^2 \frac{32m_e^2}{k^2(k^2 - 4m_e^2)^2} \int \frac{d^4q}{(2\pi)^4} \frac{1}{(p'+q)^2 - m_e^2} \frac{1}{(p+q)^2 - m_e^2}$$
$$\times \frac{1}{q^2} \left[-3k^2(p \cdot q)^2 + 2k^2 m_e^2(p \cdot q) + k^2 m_e^2 q^2 - m_e^2(k \cdot q)^2 \right]. \tag{4.2}$$

Then follow the usual steps of introducing two Feynman parameters, of performing a trivial change of variables and a symmetric integration over the
loop momentum q, so that one arrives at

$$F_2(k^2)\big|_{1 \text{ loop}} = ie^2 \frac{64m_e^2}{(k^2 - 4m_e^2)^2} \int_0^1 dx\, x \int_0^1 dy \int \frac{d^4q}{(2\pi)^4} \frac{1}{(q^2 - \mathcal{R}^2)^3}$$
$$\times \left[2x(1-x)m_e^4 - \frac{3}{4}x^2 y^2(k^2)^2 + m_e^2 k^2 x \left(3xy - y + \frac{1}{2}x \right) \right]$$
$$= \frac{e^2}{\pi^2} \frac{2m_e^2}{(k^2 - 4m_e^2)^2} \int_0^1 dx\, x \int_0^1 dy \frac{1}{\mathcal{R}^2} \times \left[2x(1-x)m_e^4 \right.$$
$$\left. - \frac{3}{4}x^2 y^2(k^2)^2 + m_e^2 k^2 x \left(3xy - y + \frac{1}{2}x \right) \right], \tag{4.3}$$

with
$$\mathcal{R}^2 = x^2 y(1 - y)(2m_e^2 - k^2) + x^2 y^2 m_e^2 + x^2(1 - y)^2 m_e^2. \tag{4.4}$$

As expected, the limit $k^2 \to 0$ can be taken without problem, and gives

$$a_e\big|_{1 \text{ loop}} \equiv F_2(0)\big|_{1 \text{ loop}} = \frac{1}{2}\frac{\alpha}{\pi}.$$

(4.5)

Let us stress that although the integral (4.1) diverges, we have obtained a finite result for $F_2(k^2)$, and hence for a_e, without introducing any regularization. This is of course expected, since a divergence in $F_2(0)$ would require that a counterterm of the form given by the second term in $\widehat{\mathcal{L}}_{\text{int}}$, see (2.8), be introduced. This would in turn spoil the renormalizability of the theory. In fact, as is well known, the divergence lies in $F_1(0)$, and goes into the renormalization of the charge of the electron.

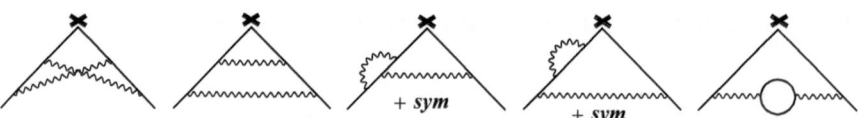

Fig. 2. The Feynman diagrams which contribute to the coefficient A_2 in (2.21).

4.2 Higher Order Mass Independent Corrections

The previous calculation is rather straightforward and amounts to the result

$$A_1 = \frac{1}{2}$$

(4.6)

first obtained by Schwinger [59]. Schwinger's calculation was soon followed by a computation of A_2 [60], which requires the evaluation of 7 graphs, representing five distinct topologies, and shown in Fig. 2. Historically, the result of Reference [60] was important, because it provided the first explicit example of the realization of the renormalization program of QED at two loops. However, the value for A_2 was not given correctly. The correct expression of the second order mass independent contribution was derived in [61–63] (see also [64,65]) and reads [10]

$$A_2 = \frac{197}{144} + \left(\frac{1}{2} - 3\ln 2\right)\zeta(2) + \frac{3}{4}\zeta(3)$$
$$= -0.328\,478\,965...$$

(4.7)

with $\zeta(p) = \sum_{n=1}^{\infty} 1/n^p$, $\zeta(2) = \pi^2/6$. The occurence of transcendental numbers like $\zeta(p)$ is a general feature of higher order calculations in perturbative

---

[10] Actually, the experimental result of [38] disagreed with the value $A_2 = -2.973$ obtained in [60], and prompted theoreticians to reconsider the calculation. The result obtained by the authors of [61–63] reconciled theory with experiment.

quantum field theory. The pattern of these transcendentals in perturbation theory and the structure of the renormalization algorithm have also been put in relationship with other mathematical structures, like knot theory and braids [66], Hopf algebras [67] and non commutative geometry [68].

The analytic evaluation of the three loop mass independent contribution to the anomalous magnetic moment required quite some time, and is mainly due to the dedication of E. Remiddi and his coworkers during the period 1969-1996. There are now 72 diagrams to consider, involving many different topologies, see Fig. 3.

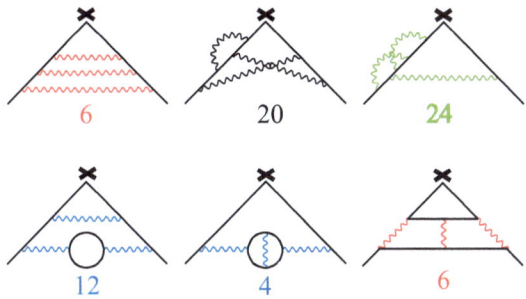

Fig. 3. The 72 Feynman diagrams which make up the coefficient A_3 in (2.21).

The calculation was completed [69] in 1996, with the analytical evaluation of a last class of diagrams, the non planar "triple cross" topologies. The result reads [11]

$$A_3 = \frac{83}{72}\pi^2\zeta(3) - \frac{215}{24}\zeta(5) + \frac{100}{3}\left[\left(a_4 + \frac{1}{24}\ln^4 2\right) - \frac{1}{24}\pi^2\ln^2 2\right]$$
$$- \frac{239}{2160}\pi^4 + \frac{139}{18}\zeta(3) - \frac{298}{9}\pi^2\ln 2 + \frac{17101}{810}\pi^2 + \frac{28259}{5184}$$
$$= 1.181\,241\,456... \tag{4.8}$$

where [12] $a_p = \displaystyle\sum_{n=1}^{\infty}\frac{1}{2^n n^p}$. The numerical value extracted from the exact analytical expression given above can be improved to any desired order of precision.

In parallel to these analytical calculations, numerical methods for the evaluation of the higher order contributions were also developed, in particular

---

[11] The completion of this three-loop program can be followed through [70]-[75] and [69]. A description of the technical aspects related to this work and an account of its status up to 1990, with references to the corresponding literature, are given in [76].

[12] The first three values are known to be $a_1 = \ln 2$, $a_2 = \text{Li}_2(1/2) = (\zeta(2) - \ln^2 2)/2$, $a_3 = \frac{7}{8}\zeta(3) - \frac{1}{2}\zeta(2)\ln 2 + \frac{1}{6}\ln^3 2$ [76].

by Kinoshita and his collaborators (for details, see [77]). The numerical evaluation of the full set of three loop diagrams was achieved in several steps
[78]-[84]. The value quoted in [84] is $A_3 = 1.195(26)$, where the error comes
from the numerical procedure. In comparison, let us quote the value [85,77]
$A_3 = 1.176\,11\,(42)$ obtained if only a subset of 21 three loop diagrams out of
the original set of 72 is evaluated numerically, relying on the analytical results
for the remaining 51 ones, and recall the value $A_3 = 1.181\,241\,456...$ obtained
from the full analytical evaluation. The error induced on a_e due to the numerical uncertainty in the second, more accurate, value is still $\Delta(a_e) = 5.3{\times}10^{-12}$,
whereas the experimental error is only $\Delta(a_e)|_{\exp} = 4.3 \times 10^{-12}$. This discus-
sion shows that the analytical evaluations of higher loop contributions to
the anomalous magnetic moment of the electron have a strong practical interest as far as the precision of the theoretical prediction is concerned, and
which goes well beyond the mere intellectual satisfaction and technical skills
involved in these calculations. [13]

At the four loop level, there are 891 diagrams to consider. Clearly, only a few
of them have been evaluated analytically [86,87]. The complete numerical
evaluation of the whole set gave [85] $A_4 = -1.434(138)$. The developement
of computers allowed subsequent reanalyses to be more accurate, i.e. $A_4 = -1.557(70)$ [88]. Until recently, the "latest of [these] constantly improving
values" was [5] $A_4 = -1.509\,8(38\,4)$. This calculation certainly represents a
formidable task, and requires many elaborate technical tools. A descriptive
account can be found in [77]. Let us mention, for completeness, that efforts
to improve upon the evaluation of A_4 are presently being pursued. Thus, a
mistake has recently been found in an earlier computer code used for the
evaluation of a subclass of four loop diagrams [89], whose contribution to A_4
was $A_{4;IV(d)} = -0.7503(60)$ [for the precise meaning of the notation, we refer
the reader to [77] and [85]]. The corrected value reads instead [89] $A_{4;IV(d)} = -0.99072(10)$ [note also the impressive improvement in the precision]. By
itself, this correction induces a 16% downward shift of the value of A_4, a far
from trivial modification [see below]. At the present stage, the value of A_4
reads

$$A_4 = -1.750\,2(38\,4). \tag{4.9}$$

Needless to say, so far the five loop contribution A_5 is unknown territory.
On the other hand, $(\alpha/\pi)^5 \sim 7 \times 10^{-14}$, so that one may reasonably expect,
in view of the present experimental situation, that its knowledge is not yet
required.

---

[13] It is only fair to point out that the numerical values that are quoted here correspond to those given in the original references. It is to be expected that they
would certainly improve if today's numerical possibilities were used.

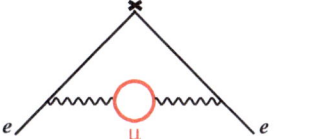 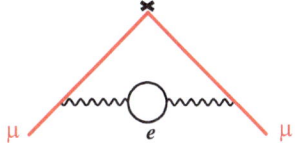

Fig. 4. The insertion of a muon vacuum polarization loop into the electron vertex correction (left) or of an electron vacuum polarization loop into the muon vertex correction (right).

4.3   Mass Dependent QED Corrections

We now turn to the QED loop contributions to the electron's anomalous magnetic moment involving the heavier leptons, μ and τ. The lowest order contribution of this type occurs at the two loop level, $\mathcal{O}(\alpha^2)$, and corresponds to a heavy lepton vacuum polarization insertion in the one loop vertex graph, cf. Fig. 4. Quite generally, the contribution to a_ℓ arising from the insertion, into the one loop vertex correction, of a vacuum polarization graph due to a loop of lepton ℓ', reads [90,91] [14]

$$B_2(\ell, \ell') = \frac{1}{3} \int_{4m_{\ell'}^2}^{\infty} ds \sqrt{1 - \frac{4m_{\ell'}^2}{s}} \, \frac{s + 2m_{\ell'}^2}{s^2} \int_0^1 dx \, \frac{x^2(1-x)}{x^2 + (1-x)\frac{s}{m_\ell^2}} . \quad (4.10)$$

If $m_{\ell'} \gg m_\ell$, the second integrand can be approximated by $x^2 m_\ell^2/s$, and one obtains [93]

$$B_2(\ell, \ell') = \frac{1}{45} \left(\frac{m_\ell}{m_{\ell'}}\right)^2 + \cdots, \quad m_{\ell'} \gg m_\ell. \quad (4.11)$$

The complete expansion of $B_2(\ell, \ell')$ for $m_{\ell'} \gg m_\ell$ can be found in [71], from which we quote

$$\begin{aligned}
B_2(\ell, \ell') = &\frac{1}{45} \left(\frac{m_\ell}{m_{\ell'}}\right)^2 + \frac{1}{70} \left(\frac{m_\ell}{m_{\ell'}}\right)^4 \ln\left(\frac{m_\ell}{m_{\ell'}}\right) + \frac{9}{19600} \left(\frac{m_\ell}{m_{\ell'}}\right)^4 \\
&+ \frac{4}{315} \left(\frac{m_\ell}{m_{\ell'}}\right)^6 \ln\left(\frac{m_\ell}{m_{\ell'}}\right) - \frac{131}{99225} \left(\frac{m_\ell}{m_{\ell'}}\right)^6 \\
&+ \mathcal{O}\left[\left(\frac{m_\ell}{m_{\ell'}}\right)^8 \ln\left(\frac{m_\ell}{m_{\ell'}}\right)\right] .
\end{aligned} \quad (4.12)$$

---

[14] A trivial change of variable on s brings the expression (4.10) into the form given in [90,91]. Furthermore, the analytical result obtained upon performing the double integration is available in [92].

Numerically, this translates into [the values for the masses that were used read $m_e = 0.510\,998\,902(21)$ MeV, $m_\tau = 1\,776.99^{+0.29}_{-0.26}$ [34], and $m_\mu/m_e = 206.768\,277(24)$ [94]]

$$B_2(e, \mu) = 5.197 \times 10^{-7}$$
$$B_2(e, \tau) = 1.838 \times 10^{-9}. \tag{4.13}$$

For later use, it is interesting to briefly discuss the structure of (4.10). The quantity which appears under the integral is related to the cross section for the scattering of a $\ell^+\ell^-$ pair into a pair $(\ell')^+(\ell')^-$ at *lowest order in QED*,

$$\sigma_{QED}^{\ell^+\ell^-\to(\ell')^+(\ell')^-}(s) = \frac{4\pi\alpha^2}{3s^2}\sqrt{1 - \frac{4m_{\ell'}^2}{s}}\,(s + 2m_{\ell'}^2), \tag{4.14}$$

so that

$$B_2(\ell; \ell') = \frac{1}{3}\int_{4m_{\ell'}^2}^{\infty}\frac{ds}{s}\,K(s)R^{(\ell')}(s), \tag{4.15}$$

where [15]

$$K(s) = \int_0^1 dx\,\frac{x^2(1 - x)}{x^2 + (1 - x)\frac{s}{m_\ell^2}}, \tag{4.16}$$

and $R^{(\ell')}(s)$ is the *lowest order* QED cross section $\sigma_{QED}^{\ell^+\ell^-\to(\ell')^+(\ell')^-}(s)$ divided by the asymptotic form of the cross section of the reaction $e^+e^- \to \mu^+\mu^-$ for $s \gg m_\mu^2$, $\sigma_\infty^{e^+e^-\to\mu^+\mu^-}(s) = \frac{4\pi\alpha^2}{3s}$.

The three loop contributions with different lepton flavours in the loops are also known analytically [95,96]. It is convenient to distinguish three classes of diagrams. The first group contains all the diagrams with one or two vacuum polarization insertion involving the same lepton, μ ot τ, of the type shown in Fig. 5. The second group consists of the leptonic light-by-light scattering insertion diagrams, Fig. 6. Finally, since there are three flavours of massive leptons in the standard model, one has also the possibility of having graphs with two heavy lepton vacuum polarization insertions, one made of a muon loop, the other of a τ loop. This gives

$$B_3(e, \ell) = B_3^{(v.p.)}(e; \mu) + B_3^{(v.p.)}(e; \tau) + B_3^{(L\times L)}(e; \mu) + B_3^{(L\times L)}(e; \tau)$$
$$+ B_3^{(v.p.)}(e; \mu, \tau). \tag{4.17}$$

The analytical expression for $B_3^{(v.p.)}(e; \mu)$ can be found in [95], whereas [96] gives the corresponding result for $B_3^{(L\times L)}(e; \mu)$. For practical purposes, it is

---

[15] Explicit expressions for $K(s)$ are also available, but for many purposes, the integral representation given here turns out to be more convenient.

both sufficient and more convenient to use their expansions in powers of m_e/m_μ,

$$
B_3^{(\text{v.p.})}(e;\mu) = \left(\frac{m_e}{m_\mu}\right)^2 \left[-\frac{23}{135}\ln\left(\frac{m_\mu}{m_e}\right) - \frac{2}{45}\pi^2 + \frac{10117}{24300}\right]
$$
$$
+ \left(\frac{m_e}{m_\mu}\right)^4 \left[\frac{19}{2520}\ln^2\left(\frac{m_\mu}{m_e}\right) - \frac{14233}{132300}\ln\left(\frac{m_\mu}{m_e}\right) + \frac{49}{768}\zeta(3)\right.
$$
$$
\left. - \frac{11}{945}\pi^2 + \frac{2976691}{296352000}\right] + \mathcal{O}\left[\left(\frac{m_e}{m_\mu}\right)^6\right]
$$
$$
= -0.000\,021\,768... \tag{4.18}
$$

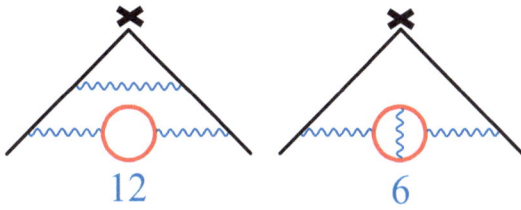

Fig. 5. Three loop QED corrections with insertion of a heavy lepton vacuum polarization which make up the coefficient $B_3^{(\text{v.p.})}(e;\mu)$.

and

$$
B_3^{(\text{L}\times\text{L})}(e;\mu) = \left(\frac{m_e}{m_\mu}\right)^2 \left[\frac{3}{2}\zeta(3) - \frac{19}{16}\right]
$$
$$
+ \left(\frac{m_e}{m_\mu}\right)^4 \left[-\frac{161}{810}\ln^2\left(\frac{m_\mu}{m_e}\right) - \frac{16189}{48600}\ln\left(\frac{m_\mu}{m_e}\right) + \frac{13}{18}\zeta(3)\right.
$$
$$
\left. - \frac{161}{9720}\pi^2 - \frac{831931}{972000}\right] + \mathcal{O}\left[\left(\frac{m_e}{m_\mu}\right)^6\right]
$$
$$
= 0.000\,014\,394\,5... \tag{4.19}
$$

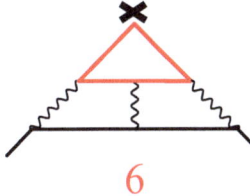

Fig. 6. The three loop QED correction with the insertion of a heavy lepton light-by-light scattering subgraph, corresponding to the coefficient $B_3^{(\text{L}\times\text{L})}(e;\mu)$.

The expressions for $B_3^{(\text{v.p.})}(e;\tau)$ and $B_3^{(\text{L}\times\text{L})}(e;\tau)$ follow upon replacing the muon mass m_μ by m_τ. This again gives a suppression factor $(m_\mu/m_\tau)^2$, which makes these contributions negligible at the present level of precision. For the same reason, $B_3^{(\text{v.p.})}(e;\mu,\tau)$ can also be discarded.

4.4   Other Contributions to a_e

In order to make the discussion of the standard model contributions to a_e complete, there remains to mention the hadronic and weak components, a_e^{had} and a_e^{weak}, respectively. Their features will be discussed in detail below, in the context of the anomalous magnetic moment of the muon. I therefore only quote the numerical values [16]

$$a_e^{\text{had}} = 1.67(3) \times 10^{-12}, \tag{4.20}$$

and [97]

$$a_e^{\text{weak}} = 0.030 \times 10^{-12}. \tag{4.21}$$

4.5   Comparison with Experiment and Determination of α

Summing up the various contributions discussed so far gives the standard model prediction [4,5,8,89]

$$a_e^{\text{SM}} = 0.5\frac{\alpha}{\pi} - 0.328\,478\,444\,00\left(\frac{\alpha}{\pi}\right)^2 + 1.181\,234\,017\left(\frac{\alpha}{\pi}\right)^3$$
$$- 1.750\,2(38\,4)\left(\frac{\alpha}{\pi}\right)^4 + 1.70(3) \times 10^{-12}. \tag{4.22}$$

In order to obtain a number that can be compared to the experimental result, a sufficiently accurate determination of the fine structure constant α is required. The best available measurement of the latter comes from the quantum Hall effect [33],

$$\alpha^{-1}(qH) = 137.036\,003\,00(2\,70) \tag{4.23}$$

and leads to

$$a_e^{\text{SM}}(qH) = 0.001\,159\,652\,146\,5(24\,0), \tag{4.24}$$

about six times less accurate than the experimental value of (3.2), $a_e^{\text{exp}} = 0.001\,159\,652\,188\,3(4\,2)$. On the other hand, if one excludes other contributions to a_e than those from the standard model considered so far, i.e. if one

---

[16] I reproduce here the values given in [4,5], except for the fact that I have taken into account the changes in the value of the hadronic light-by-light contribution to a_μ, see below, for which I take $a_\mu^{(\text{L}\times\text{L})} = +8(4) \times 10^{-10}$, and which translates into $a_e^{(\text{L}\times\text{L})} \sim a_\mu^{(\text{L}\times\text{L})}(m_e/m_\mu)^2 = 0.02 \times 10^{-12}$.

identifies a_e^{SM} with a_e^{exp}, then the value of a_e^{exp} as given in (3.2) provides the best determination of α to date [89,8],

$$\alpha^{-1}(a_e) = 137.035\,998\,75(52)\,, \tag{4.25}$$

at least to the extend that one may reasonably believe that all theoretical errors are under control. Now, as we have seen earlier, the value of A_4 has recently been corrected [89]. The analysis presented here incorporates the changes brought forward by the analysis of [89]. It lowers the prediction for a_e by $\sim 7.0 \times 10^{-12}$ if one uses the value (4.23) of α, or equivalently reduces the value of $\alpha^{-1}(a_e)$ by one and a half standard deviation. It is very likely that the completion of the analysis begun in [89] will lead to a more accurate determination of A_4, and further changes in the numbers quoted here are to be expected.

Exercises for Section 4

Exercise 4.1
Reproduce the steps that lead from (4.1) to (4.5).

Exercise 4.2
Consider the contribution to a_ℓ induced by the vacuum polarization due to the loop of a fermion ℓ', as shown in Fig. 4. Show that the corresponding contribution to the matrix element (2.4) reads

$$\Gamma^\rho(p',p)\big|_{2\,\text{loop};(\ell;\ell')} = (-ie_0)^2 \int \frac{d^4q}{(2\pi)^4} \gamma^\mu \frac{i}{\not{p}'+\not{q}-m_e} \gamma^\rho \frac{i}{\not{p}+\not{q}-m_e} \gamma^\nu$$
$$\times \frac{(-i)}{q^2}\left[\eta_{\mu\lambda} - (1-\xi)\frac{q_\mu q_\lambda}{q^2}\right]$$
$$\times \frac{(-i)}{q^2}\left[\eta_{\nu\sigma} - (1-\xi)\frac{q_\nu q_\sigma}{q^2}\right]$$
$$\times (-ie_0)^2 \frac{1}{i} \Pi_{(\ell')}^{\lambda\sigma}(q^2)\,,$$

with

$$\Pi_{(\ell')}^{\mu\nu}(q^2) = i \int d^4x e^{iq\cdot x} \langle \Omega| \text{T}\{j_{(\ell')}^\mu(x) j_{(\ell')}^\nu(0)\}|\Omega\rangle\,,$$

and $j_{(\ell')}^\mu(x) = \overline{\psi}_{(\ell')} \gamma^\mu \psi_{(\ell')}$ is the (conserved) elctromagnetic current of the fermion ℓ'. Finally, e_0 denotes the bare electric charge. Show that

$$\Pi_{(\ell')}^{\mu\nu}(q^2) = (q^\mu q^\nu - \eta^{\mu\nu}q^2)\,\Pi_{(\ell')}(q^2)$$

and that

$$\Gamma^\rho(p',p)\big|_{2\,\text{loop};(\ell;\ell')} = (-ie_0)^2 \int \frac{d^4q}{(2\pi)^4} \gamma^\mu \frac{i}{\not{p}'+\not{q}-m_e} \gamma^\rho \frac{i}{\not{p}+\not{q}-m_e} \gamma^\nu$$

$$\times (-ie_0)^2 \frac{1}{i} \Pi_{(\ell')}(q^2) \left(\frac{-i}{q^2}\right)^2 (q_\mu q_\nu - \eta_{\mu\nu}q^2) \,.$$

Show that the part proportional to $q_\mu q_\nu$ in the last term contributes only to $F_1(q^2)$. The function $\Pi_{(\ell')}(q^2)$ is divergent. The divergence is contained in $\Pi_{(\ell')}(0)$, so that one may write

$$\Pi_{(\ell')}(q^2) = \Pi_{(\ell')}(0) + \Pi^{\mathrm{ren}}_{(\ell')}(q^2)$$

where the renormalized function $\Pi^{\mathrm{ren}}_{(\ell')}(q^2)$ obeys a dispersion relation

$$\Pi^{\mathrm{ren}}_{(\ell')}(q^2) = q^2 \int \frac{dt}{t} \frac{1}{t-q^2} \frac{1}{\pi} Im\Pi_{(\ell')}(t) \,.$$

Show that the contribution from the divergent part $\Pi_{(\ell')}(0)$ to $F_2(k^2)$ writes in the form

$$F_2(k^2)\big|_{1 \text{ loop}} \times e_0^2 \, (-1) \, \Pi_{(\ell')}(0) \,,$$

and that it therefore corresponds to a renormalization, due to the ℓ' fermion loop, of the bare electric charge e_0,

$$e = e_0 \sqrt{Z_3} \,, \quad Z_3^{(\ell')}\big|_{1 \text{ loop}} = \frac{1}{1 + e^2 \Pi_{(\ell')}(0)} \sim 1 - e^2 \Pi_{(\ell')}(0) \,,$$

that appears in the one loop result $F_2(k^2)\big|_{1 \text{ loop}}$. Finally, show that the renormalized component $\Pi^{\mathrm{ren}}_{(\ell')}(q^2)$ generates a contribution to a_ℓ that can be written as

$$a_\ell\big|_{2 \text{ loop};(\ell;\ell')} = \int \frac{dt}{t} \frac{1}{\pi} Im\Pi_{(\ell')}(t) \times a_\ell^{1 \text{ loop}}(t) \,,$$

where $a_\ell^{1 \text{ loop}}(t)$ denotes the expression (4.2) for $k^2 = 0$ and with the denominator $\mathcal{R}$ replaced by $\mathcal{R}(t) = \mathcal{R} - (1-x)t$.

5   The Anomalous Magnetic Moment of the Muon

In this section, the theoretical aspects concerning the anomalous magnetic moment of the muon are discussed. Since the muon is much heavier than the electron, a_μ will be more sensitive to higher mass scales. In particular, it is a better probe for possible degrees of freedom beyond the standard model, like supersymmetry. The drawback, however, is that a_μ will also be more sensitive to the non perturbative strong interaction dynamics at the ~ 1 GeV scale.

5.1   QED Contributions to a_μ

As already mentioned before, the mass independent QED contributions to a_μ are described by the same coefficients A_n as in the case of the electron. We therefore need only to discuss the coefficients $B_n(\mu; \ell')$ associated with the mass dependent corrections.

For $m_{\ell'} \ll m_\ell$, (4.10) gives [90–92,71]

$$B_2(\ell; \ell') = \frac{1}{3} \ln\left(\frac{m_\ell}{m_{\ell'}}\right) - \frac{25}{36} + \frac{\pi^2}{4} \frac{m_{\ell'}}{m_\ell} - 4 \left(\frac{m_{\ell'}}{m_\ell}\right)^2 \ln\left(\frac{m_\ell}{m_{\ell'}}\right)$$
$$+ 3\left(\frac{m_{\ell'}}{m_\ell}\right)^2 + \mathcal{O}\left[\left(\frac{m_{\ell'}}{m_\ell}\right)^3\right]. \tag{5.1}$$

The complete expansion in powers of $m_{\ell'}/m_\ell$ can again be found in [71]. In the case of $B_2(\mu; \tau)$, one may use the expression given in (4.12). Upon using the values $m_e = 0.510\,998\,902(21)$ MeV, $m_\mu = 105.658\,357(5)$ MeV, $m_\tau = 1776.99^{+0.29}_{-0.26}$ MeV [34], and $m_\mu/m_e = 206.768\,277(24)$ [94] the corresponding numbers read

$$B_2(\mu; e) = 1.094\,258\,300(38) \tag{5.2}$$
$$B_2(\mu; \tau) = 0.000\,078\,064(25). \tag{5.3}$$

Although these numbers follow from an analytical expression, there are uncertainties attached to them, induced by those on the corresponding values of the ratios of the lepton masses.

The three loop QED corrections decompose as

$$B_3(\mu, \ell) = B_3^{(\text{v.p.})}(\mu; e) + B_3^{(\text{v.p.})}(\mu; \tau) + B_3^{(\text{L}\times\text{L})}(\mu; e) + B_3^{(\text{L}\times\text{L})}(\mu; \tau)$$
$$+ B_3^{(\text{v.p.})}(\mu; e, \tau), \tag{5.4}$$

with [95,96]

$$B_3^{(\text{v.p.})}(\mu; e) = \frac{2}{9} \ln^2\left(\frac{m_\mu}{m_e}\right) + \left[\zeta(3) - \frac{2}{3}\pi^2 \ln 2 + \frac{1}{9}\pi^2 + \frac{31}{27}\right] \ln\left(\frac{m_\mu}{m_e}\right)$$
$$+ \frac{11}{216}\pi^4 - \frac{2}{9}\pi^2 \ln^2 2 - \frac{8}{3}a_4 - \frac{1}{9}\ln^4 2 - 3\zeta(3) + \frac{5}{3}\pi^2 \ln 2 - \frac{25}{18}\pi^2$$
$$+ \frac{1075}{216} + \frac{m_e}{m_\mu}\left[-\frac{13}{18}\pi^3 - \frac{16}{9}\pi^2 \ln 2 + \frac{3199}{1080}\pi^2\right]$$
$$+ \left(\frac{m_e}{m_\mu}\right)^2 \left[\frac{10}{3} \ln^2\left(\frac{m_\mu}{m_e}\right) - \frac{11}{9} \ln\left(\frac{m_\mu}{m_e}\right) - \frac{14}{3}\pi^2 \ln 2 - 2\zeta(3) + \frac{49}{12}\pi^2\right.$$
$$\left. - \frac{131}{54}\right] + \left(\frac{m_e}{m_\mu}\right)^3 \left[\frac{4}{3}\pi^2 \ln\left(\frac{m_\mu}{m_e}\right) + \frac{35}{12}\pi^3 - \frac{16}{3}\pi^2 \ln 2 - \frac{5771}{1080}\pi^2\right]$$

$$+ \left(\frac{m_e}{m_\mu} \right)^4 \left[-\frac{25}{9} \ln^3 \left(\frac{m_\mu}{m_e} \right) - \frac{1369}{180} \ln^2 \left(\frac{m_\mu}{m_e} \right) \right.$$

$$+ [-2\zeta(3) + 4\pi^2 \ln 2 - \frac{269}{144} \pi^2 - \frac{7496}{675}] \ln \left(\frac{m_\mu}{m_e} \right)$$

$$- \frac{43}{108} \pi^4 + \frac{8}{9} \pi^2 \ln^2 2 + \frac{80}{3} a_4 + \frac{10}{9} \ln^4 2 - \frac{411}{32} \zeta(3) + \frac{89}{48} \pi^2 \ln 2$$

$$\left. - \frac{1061}{864} \pi^2 - \frac{274511}{54000} \right] + \mathcal{O} \left[\left(\frac{m_e}{m_\mu} \right)^5 \right], \tag{5.5}$$

$$B_3^{(\mathrm{L} \times \mathrm{L})}(\mu; e) = \frac{2}{3} \pi^2 \ln \left(\frac{m_\mu}{m_e} \right) + \frac{59}{270} \pi^4 - 3\zeta(3) - \frac{10}{3} \pi^2 + \frac{2}{3}$$

$$+ \frac{m_e}{m_\mu} \left[\frac{4}{3} \pi^2 \ln \left(\frac{m_\mu}{m_e} \right) - \frac{196}{3} \pi^2 \ln 2 + \frac{424}{9} \pi^2 \right]$$

$$+ \left(\frac{m_e}{m_\mu} \right)^2 \left[-\frac{2}{3} \ln^3 \left(\frac{m_\mu}{m_e} \right) + (\frac{\pi^2}{9} - \frac{20}{3}) \ln^2 \left(\frac{m_\mu}{m_e} \right) \right.$$

$$- [\frac{16}{135} \pi^4 + 4\zeta(3) - \frac{32}{9} \pi^2 + \frac{61}{3}] \ln \left(\frac{m_\mu}{m_e} \right)$$

$$\left. + \frac{4}{3} \zeta(3) \pi^2 - \frac{61}{270} \pi^4 + 3\zeta(3) + \frac{25}{18} \pi^2 - \frac{283}{12} \right]$$

$$+ \left(\frac{m_e}{m_\mu} \right)^3 \left[\frac{10}{9} \pi^2 \ln \left(\frac{m_\mu}{m_e} \right) - \frac{11}{9} \pi^2 \right]$$

$$+ \left(\frac{m_e}{m_\mu} \right)^4 \left[\frac{7}{9} \ln^3 \left(\frac{m_\mu}{m_e} \right) + \frac{41}{18} \ln^2 \left(\frac{m_\mu}{m_e} \right) + \frac{13}{9} \pi^2 \ln \left(\frac{m_\mu}{m_e} \right) \right.$$

$$\left. + \frac{517}{108} \ln \left(\frac{m_\mu}{m_e} \right) + \frac{1}{2} \zeta(3) + \frac{191}{216} \pi^2 + \frac{13283}{2592} \right] + \mathcal{O} \left[\left(\frac{m_e}{m_\mu} \right)^5 \right], \tag{5.6}$$

while $B_3^{(\mathrm{v.p.})}(\mu; \tau)$ and $B_3^{(\mathrm{L} \times \mathrm{L})}(\mu; \tau)$ are derived from $B_3^{(\mathrm{v.p.})}(e; \mu)$ and from $B_3^{(\mathrm{L} \times \mathrm{L})}(e; \mu)$, respectively, by trivial substitutions of the masses. Furthermore, the graphs with mixed vacuum polarization insertions, one electron loop, and one τ loop, are evaluated using a dispersive integral [71,95,98].

Numerically, one obtains

$$B_3^{(\mathrm{v.p.})}(\mu; e) = 1.920\,455\,1(2)$$
$$B_3^{(\mathrm{L} \times \mathrm{L})}(\mu; e) = 20.947\,924\,7(7)$$
$$B_3^{(\mathrm{v.p.})}(\mu; \tau) = -0.001\,782\,3(5)$$
$$B_3^{(\mathrm{L} \times \mathrm{L})}(\mu; \tau) = 0.002\,142\,9(7)$$
$$B_3^{(\mathrm{v.p.})}(\mu; e, \tau) = 0.000\,527\,7(2) \,. \tag{5.7}$$

Notice the large value of $B_3^{(\mathrm{L \times L})}(\mu; e)$, due to the occurence of terms involving factors like $\ln(m_\mu/m_e) \sim 5$ and powers of π. Such a large contribution, first obtained numerically in [78], allowed to explain a discrepancy of $1.7\,\sigma$ between the theoretical value and the experimental measurement of [45]. Finally, several pieces of $B_3^{(\mathrm{v.p.})}(\mu; e)$ had already been worked out earlier, in [24,93,99–101]

The contributions at fourth order in α have been obtained numerically in [102]. They must be corrected for the change in A_4 obtained in [89]. At the next order, no full calculation, even through numerical techniques, is available. Specific contributions, for instance those enhanced by powers of $\ln(m_e/m_\mu)$ times powers of π, have been evaluated [103–106].

Putting all these contributions together leads to the expression

$$a_\mu^{\mathrm{QED}} = 0.5\,\frac{\alpha}{\pi} + 0.765\,857\,399(45)\left(\frac{\alpha}{\pi}\right)^2 + 24.050\,509\,5(2\,3)\left(\frac{\alpha}{\pi}\right)^3$$
$$+ 125.08(41)\left(\frac{\alpha}{\pi}\right)^4 + 930(170)\left(\frac{\alpha}{\pi}\right)^5. \tag{5.8}$$

Upon inserting the value of α obtained from the anomalous magnetic moment of the electron in (4.25), one finds

$$a_\mu^{\mathrm{QED}} = 11\,658\,470.35(28) \times 10^{-10}. \tag{5.9}$$

5.2   Hadronic Contributions to a_μ

On the level of Feynman diagrams, hadronic contributions arise through loops of virtual quarks and gluons. These loops also involve the soft scales, and therefore cannot be computed reliably in perturbative QCD. We shall decompose the hadronic contributions into three subsets: hadronic vacuum polarization insertions at order α^2, at order α^3, and hadronic light-by-light scattering,

$$a_\mu^{\mathrm{had}} = a_\mu^{(\mathrm{h.v.p.\ 1})} + a_\mu^{(\mathrm{h.v.p.\ 2})} + a_\mu^{(\mathrm{h.\ L \times L})} \tag{5.10}$$

Hadronic Vacuum Polarization. We first discuss $a_\mu^{(\mathrm{h.v.p.\ 1})}$, which arises at order $\mathcal{O}(\alpha^2)$ from the insertion of a single hadronic vacuum polarization into the lowest order vertex correction graph, see Fig. 7. The importance of this contribution to a_μ is known since long time [107,108].

There is a very convenient dispersive representation of this diagram, similar to (4.10)

$$a_\mu^{(\mathrm{h.v.p.\ 1})} = 4\alpha^2 \int_{4M_\pi^2}^\infty \frac{ds}{s} K(s) \frac{1}{\pi} \mathrm{Im}\Pi(s)$$
$$= \frac{1}{3}\left(\frac{\alpha}{\pi}\right)^2 \int_{4M_\pi^2}^\infty \frac{ds}{s} K(s) R^{\mathrm{had}}(s), \tag{5.11}$$

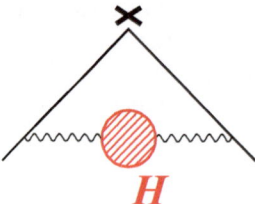

Fig. 7. The insertion of the hadronic vacuum polarization into the one loop vertex correction, corresponding to $a_\mu^{(\text{h.v.p. 1})}$.

Here, $\Pi(s)$ denotes the *hadronic* vacuum polarization function, defined as [17]

$$(q_\mu q_\nu - q^2 \eta_{\mu\nu})\Pi(Q^2) = i \int d^4x e^{iq\cdot x} \langle \Omega | T\{j_\mu(x)j_\nu(0)\}|\Omega\rangle, \qquad (5.12)$$

with j_ρ the hadronic component of the electromagnetic current, $Q^2 = -q^2 \geq 0$ for q_μ spacelike, and $|\Omega\rangle$ the QCD vacuum. The function $K(s)$ was defined in (4.16), and $R^{\text{had}}(s)$ stands now [see however below] for the cross section $\sigma_0^{e^+e^- \to \text{had}}(s)$ of $e^+e^- \to$ hadrons, *at lowest order in* α, divided by $\sigma_\infty^{e^+e^- \to \mu^+\mu^-}(s) = \frac{4\pi\alpha^2}{3s}$. A first principle computation of this strong interaction contribution is far beyond our present abilities to deal with the non perturbative aspects of confining gauge theories. This last relation is however very interesting because it expresses $a_\mu^{(\text{h.v.p. 1})}$ through a quantity that can be measured experimentally. In this respect, two important properties of the function $K(s)$ deserve to be mentioned. First, it appears from the integral representation (4.16) that $K(s)$ is positive definite. Since R^{had} is also positive, one deduces that $a_\mu^{(\text{h.v.p. 1})}$ itself is positive. Second, the function $K(s)$ decreases as $m_\mu^2/3s$ as s grows, so that it is indeed the low energy region which dominates the integral. Explicit evaluation of $a_\mu^{(\text{h.v.p. 1})}$ using available data actually reveals that more than 80% of its value comes from energies below 1.4 GeV. This observation is rather welcome, since the spectral density $\text{Im}\Pi(s)$ can also be extracted from data on the hadronic decays of the τ lepton in this energy region.

Finally, the values obtained in this way for $a_\mu^{(\text{h.v.p. 1})}$ have evolved in time, as shown in Table 3. This evolution is mainly driven by the availability of more data, and is still going on, as the last entries of Table 3 show. In order to match the precision reached by the latest experimental measurement of a_μ, $a_\mu^{(\text{h.v.p. 1})}$ needs to be known at $\sim 1\%$. Besides the very recent high quality e^+e^- data obtained by the BES Collaboration [109] in the region between 2 to 5 GeV, and by the CMD-2 collaboration [110] in the region dominated by

---

[17] Actually, $\Pi(s)$ defined this way has an ultraviolet divergence, produced by the QCD short distance singularity of the chronological product of the two currents. However, it only affects the real part of $\Pi(s)$. A renormalized, finite quantity is obtained by a single subtraction, $\Pi(s) - \Pi(0)$.

Table 3. Some of the recent evaluations of $a_\mu^{(\text{h.v.p. 1})} \times 10^{11}$ from e^+e^- and/or τ-decay data. Note that the authors of [120] also use space like data for the pion form factor $F_\pi(t)$; furthermore, they use a preliminary version of the CMD-2 data, that is now superseded by [110].

7024(153)	[115] e^+e^-
7026(160)	[116] e^+e^-
6950(150)	[117] e^+e^-
7011(94)	[117] τ, e^+e^-,
6951(75)	[118] τ, e^+e^-, QCD
6924(62)	[119] τ, e^+e^-, QCD
7016(119)	[58] e^+e^-, QCD
7036(76)	[58] τ, e^+e^-, QCD
7002(73)	[120] e^+e^-, incl. BES-II data, $F_\pi(t)$
6836(86)	[121] e^+e^-, incl. BES-II and CMD-2 data
6847(70)	[122] e^+e^-, incl. BES-II and CMD-2 data
7090(59)	[122] τ, e^+e^-, incl. BES-II data
6831(62)	[123] e^+e^-, incl. BES-II and CMD-2 data

the ρ resonance, the latest analyses sometimes also include or use, in the low-energy region, data obtained from hadronic decays of the τ by ALEPH [111], and, more recently, by OPAL [112] and CLEO [113,114]. We may notice from Table 3 that the precision obtained by using e^+e^- data alone has become comparable to the one achieved upon including the τ data. However, one of the latest analyses reveals a troubling discrepancy between the purely e^+e^- and the τ based evaluations. Additional work is certainly needed in order to resolve these problems [18]. Further data are also expected in the future, from the KLOE experiment at the DAPHNE e^+e^- machine [124], or from the B factories BaBar [125] and Belle. For additional comparative discussions and details of the various analyses, we refer the reader to the literature quoted in Table 3. Averaging the results from the three most recent analysis with e^+e^- data only, from [121], [122], and [123], gives [8]

$$a_\mu^{(\text{h.v.p. 1})}(e^+e^-) = 6838(75) \times 10^{-11}. \tag{5.13}$$

Let us briefly mention here that it is quite easy to estimate the order of magnitude of $a_\mu^{(\text{h.v.p. 1})}$. For this purpose, it is convenient to introduce still another representation [2,126], which relates $a_\mu^{(\text{h.v.p. 1})}$ to the hadronic Adler function $\mathcal{A}(Q^2)$, defined as [19]

$$\mathcal{A}(Q^2) = -Q^2 \frac{\partial \Pi(Q^2)}{\partial Q^2} = \int_0^\infty dt \, \frac{Q^2}{(t+Q^2)^2} \frac{1}{\pi} \text{Im}\Pi(t), \tag{5.14}$$

---

[18] For a possible explanation, see [8].
[19] Unlike $\Pi(t)$ itself, $\mathcal{A}(Q^2)$ if free from ultraviolet divergences.

by

$$a_\mu^{(\text{h.v.p. 1})} = 2\pi^2 \left(\frac{\alpha}{\pi}\right)^2 \int_0^1 \frac{dx}{x} (1-x)(2-x)\mathcal{A}\left(\frac{x^2}{1-x}m_\mu^2\right). \qquad (5.15)$$

A simple representation of the hadronic Adler function can be obtained if one assumes that $\text{Im}\Pi(t)$ is given by a single, zero width, vector meson pole, and, above a certain threshold s_0, by the QCD perturbative continuum contribution,

$$\frac{1}{\pi}\text{Im}\Pi(t) = \frac{2}{3} f_V^2 M_V^2 \delta(t - M_V^2) + \frac{2}{3}\frac{N_C}{12\pi^2}\left[1 + \mathcal{O}(\alpha_s)\right]\theta(t - s_0) \qquad (5.16)$$

The justification [127] for this type of minimal hadronic ansatz can be found within the framework of the large-N_C limit [128,129] of QCD, see [127] for a general discussion and a detailed study of this representation of the Adler function. The threshold s_0 for the onset of the continuum can be fixed from the property that in QCD there is no contribution in $1/Q^2$ in the short distance expansion of $\mathcal{A}(Q^2)$, which requires [127]

$$2f_V^2 M_V^2 = \frac{N_C}{12\pi^2} s_0 \left(1 + \frac{3}{8}\frac{\alpha_s(s_0)}{\pi} + \mathcal{O}(\alpha_s^2)\right). \qquad (5.17)$$

This then gives $a_\mu^{(\text{h.v.p. 1})} \sim (570\pm170)\times10^{-10}$, which compares well with the more elaborate data based evaluations in Table 3, even though this simple estimate cannot claim to provide the required accuracy of about 1%.

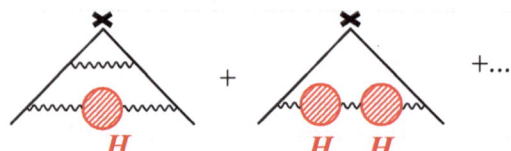

Fig. 8. Higher order corrections containing the hadronic vacuum polarization contribution, corresponding to $a_\mu^{(\text{h.v.p. 2})}$.

We now come to the $\mathcal{O}(\alpha^3)$ corrections involving hadronic vacuum polarization subgraphs. Besides the contributions shown in Fig. 8, another one is obtained upon inserting a lepton loop in one of the two photon lines of the graph shown in Fig. 7. Taken all together, these can again be expressed in terms of R^{had} [130,3,98]

$$a_\mu^{(\text{h.v.p. 2})} = \frac{1}{3}\left(\frac{\alpha}{\pi}\right)^3 \int_{4M_\pi^2}^\infty \frac{ds}{s} K^{(2)}(s)R^{\text{had}}(s). \qquad (5.18)$$

The value obtained for this quantity is [98]

$$a_\mu^{(\text{h.v.p. 2})} \times 10^{11} = -101 \pm 6. \qquad (5.19)$$

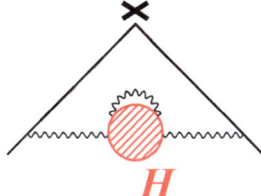

Fig. 9. A higher order correction containing the hadronic vacuum polarization contribution, and which is included in $a_\mu^{(\text{h.v.p. 1})}$.

The expression of $K^{(2)}(s)$ is given in the references quoted above.

There is actually another $\mathcal{O}(\alpha^3)$ correction, namely the one obtained upon attaching a virtual photon line with both ends to the hadronic blob in Fig. 7, see Fig. 9. On the other hand, $a_\mu^{(\text{h.v.p. 1})}$ involves in principle data corrected for *all* electromagnetic effects. Whereas radiative corrections in the leptonic initial state and vacuum polarization effects in the photon propagator certainly can be accounted for, there is at present no way to handle in a model independent way electromagnetic corrections in the hadronic final state. These, on the other hand, contribute, together with final states containing an additional photon, to the $\mathcal{O}(\alpha^3)$ contribution we have just been mentioning. It has become customary to include all these effects into $a_\mu^{(\text{h.v.p. 1})}$, where it is then to be understood that $R^{\text{had}}(s)$ in (5.11) actually stands for

$$R^{\text{had}}(s) = \frac{\sigma_0^{e^+e^-\to\text{had}}(s) + \sigma_2^{e^+e^-\to\text{had}}(s) + \sigma_0^{e^+e^-\to\text{had}+\gamma}(s)}{\sigma_\infty^{e^+e^-\to\mu^+\mu^-}(s)}, \quad (5.20)$$

where $\sigma_0^{e^+e^-\to\text{had}}(s) + \sigma_2^{e^+e^-\to\text{had}}(s)$ denotes the cross section for $e^+e^- \to$ had beyond leading order in the expansion in powers of the fine structure constant α. The values given in Table 3 correspond to the definition (5.20). It also should be stressed that the next-to-leading cross section $\sigma_2^{e^+e^-\to\text{had}}(s)$, as well as the radiative cross section $\sigma_0^{e^+e^-\to\text{had}+\gamma}(s)$, are infrared divergent quantities, and only their sum is actually well defined.

Hadronic Light-by-Light Scattering. We now discuss the so called hadronic light-by-light scattering graphs of Fig. 10.

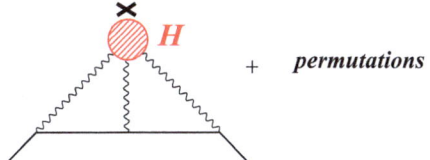

$+$ *permutations*

Fig. 10. The hadronic light-by-light scattering graphs contributing to $a_\mu^{(\text{h. L}\times\text{L})}$.

The contribution to $\Gamma_\rho(p',p)$ of relevance here is the matrix element, at lowest nonvanishing order in the fine structure constant α, of the light quark electromagnetic current

$$j_\rho(x) = \frac{2}{3}(\bar{u}\gamma_\rho u)(x) - \frac{1}{3}(\bar{d}\gamma_\rho d)(x) - \frac{1}{3}(\bar{s}\gamma_\rho s)(x) \tag{5.21}$$

between μ^- states,

$$\begin{aligned}
(-ie)\bar{u}(p')&\Gamma_\rho^{(\text{h. L}\times\text{L})}(p',p)u(p) \equiv \langle \mu^-(p')|(ie)j_\rho(0)|\mu^-(p)\rangle \\
&= \int \frac{d^4q_1}{(2\pi)^4} \int \frac{d^4q_2}{(2\pi)^4} \frac{(-i)^3}{q_1^2\, q_2^2\,(q_1+q_2-k)^2} \\
&\times \frac{i}{(p'-q_1)^2 - m^2} \frac{i}{(p'-q_1-q_2)^2 - m^2} \\
&\times (-ie)^3 \bar{u}(p')\gamma^\mu(\slashed{p}'-\slashed{q}_1+m)\gamma^\nu(\slashed{p}'-\slashed{q}_1-\slashed{q}_2+m)\gamma^\lambda u(p) \\
&\times (ie)^4 \Pi_{\mu\nu\lambda\rho}(q_1,q_2,k-q_1-q_2)\,,
\end{aligned} \tag{5.22}$$

with $k_\mu = (p'-p)_\mu$ and

$$\begin{aligned}
\Pi_{\mu\nu\lambda\rho}(q_1,q_2,q_3) = \int d^4x_1 \int d^4x_2 \int d^4x_3\, e^{i(q_1\cdot x_1 + q_2\cdot x_2 + q_3\cdot x_3)} \\
\times \langle \Omega\,|\,\text{T}\{j_\mu(x_1)j_\nu(x_2)j_\lambda(x_3)j_\rho(0)\}\,|\,\Omega\rangle
\end{aligned} \tag{5.23}$$

the fourth-rank light quark hadronic vacuum-polarization tensor, $|\,\Omega\rangle$ denoting the QCD vacuum. Since the flavour diagonal current $j_\mu(x)$ is conserved, the tensor $\Pi_{\mu\nu\lambda\rho}(q_1,q_2,q_3)$ satisfies the Ward identities

$$\{q_1^\mu; q_2^\nu; q_3^\lambda; (q_1+q_2+q_3)^\rho\}\Pi_{\mu\nu\lambda\rho}(q_1,q_2,q_3) = \{0;0;0;0\}\,. \tag{5.24}$$

This entails that [20] $\Gamma_\rho^{(\text{h. L}\times\text{L})}(p',p) = k^\tau \Gamma_{\rho\tau}^{(\text{h. L}\times\text{L})}(p',p)$, with

$$\begin{aligned}
\bar{u}(p')\Gamma_{\rho\sigma}^{(\text{h. L}\times\text{L})}(p',p)u(p) &= -ie^6 \int \frac{d^4q_1}{(2\pi)^4} \int \frac{d^4q_2}{(2\pi)^4} \frac{1}{q_1^2\, q_2^2\,(q_1+q_2-k)^2} \\
&\times \frac{1}{(p'-q_1)^2 - m^2} \frac{1}{(p'-q_1-q_2)^2 - m^2} \\
&\times \bar{u}(p')\gamma^\mu(\slashed{p}'-\slashed{q}_1+m)\gamma^\nu(\slashed{p}'-\slashed{q}_1-\slashed{q}_2+m)\gamma^\lambda u(p) \\
&\times \frac{\partial}{\partial k^\rho}\Pi_{\mu\nu\lambda\sigma}(q_1,q_2,k-q_1-q_2)\,.
\end{aligned} \tag{5.25}$$

Following [78] and using the property $k^\rho k^\sigma \bar{u}(p')\Gamma_{\rho\sigma}^{(\text{h. L}\times\text{L})}(p',p)u(p) = 0$, one deduces that the hadronic light-by-light contribution to the muon anomalous

---

[20] We use the following conventions for Dirac's γ-matrices: $\{\gamma_\mu,\gamma_\nu\} = 2\eta_{\mu\nu}$, with $\eta_{\mu\nu}$ the flat Minkowski space metric of signature $(+---)$, $\sigma_{\mu\nu} = (i/2)[\gamma_\mu,\gamma_\nu]$, $\gamma_5 = i\gamma^0\gamma^1\gamma^2\gamma^3$, whereas the totally antisymmetric tensor $\varepsilon_{\mu\nu\rho\sigma}$ is chosen such that $\varepsilon_{0123} = +1$.

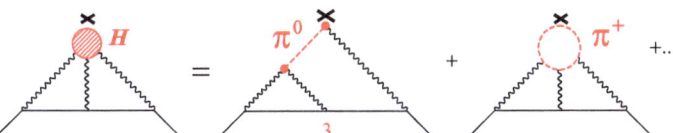

Fig. 11. Some individual contributions to hadronic light-by-light scattering: the neutral pion exchange and the charged pion loop. There are other contributions, not shown here.

magnetic moment is equal to

$$a_\mu^{(\text{h. L×L})} = \frac{1}{48m} \, \text{tr} \left\{ (\not{p}+m)[\gamma^\rho, \gamma^\sigma](\not{p}+m)\Gamma_{\rho\sigma}^{(\text{h. L×L})}(p,p) \right\} . \qquad (5.26)$$

This is about all we can say on general grounds about the QCD four-point function $\Pi_{\mu\nu\lambda\rho}(q_1, q_2, q_3)$. Unlike the hadronic vacuum polarization function, there is no experimental data which would allow for an evaluation of $a_\mu^{(\text{h. L×L})}$. The existing estimates regarding this quantity therefore rely on specific models in order to account for the non perturbative QCD aspects. A few particular contributions can be identified, see Fig. 11. For instance, there is a contribution where the four photon lines are attached to a closed loop of charged mesons. The case of the charged pion loop with pointlike couplings is actually finite and contributes $\sim 4 \times 10^{-10}$ to a_μ [131]. If the coupling of charged pions to photons is modified by taking into account the effects of resonances like the ρ, this contribution is reduced by a factor varying between 3 [131,132] and 10 [133], depending on the resonance model used. Another class of contributions consists of those involving resonance exchanges between photon pairs [131–134]. Although here also the results depend on the models used, there is a constant feature that emerges from all the analyses that have been done: the contribution coming from the exchange of the pseudoscalars, π^0, η and η' gives practically the final result. Other contributions [charged pion loops, vector, scalar, and axial resonances,...] are not only smaller, but also tend to cancel among themselves.

Some of the results obtained for $a_\mu^{(\text{h. L×L})}$ have been collected in Table 4. Leaving aside the first result [130,3] shown there, which is affected by a bad numerical convergence [131], one notices that the sign of this contribution has changed twice. The first change resulted from a mistake in [131], that was corrected for in [133]. The minus sign that resulted was confirmed by an independent calculation, using the ENJL model, in [132]. A subsequent reanalysis [134] gave additional support to a negative result, while also getting better agreement with the value of [132].

Needless to say, these evaluations are based on heavy numerical work, which has the drawback of making the final results rather opaque to an intuitive understanding of the physics behind them. We [135] therefore decided to improve things on the analytical side, in order to achieve a better understanding of the relevant features that led to the previous results. Taking advantage of

Table 4. Various evaluations of $a_\mu^{(\text{h. L}\times\text{L})} \times 10^{11}$ (*first column*) and of the neutral pion exchange contribution $a_\mu^{(\text{h. L}\times\text{L};\pi^0)} \times 10^{11}$ (*second column*)

$-260(100)$		constituent quark loop	[130,3]
$+60(4)$		constituent quark loop	[131]
$+49(5)$	$+65(6)$	$\pi^\pm$loop, π^0 and resonance exchanges	[131]
$-92(32)$		ENJL, $a_\mu^{(\text{h. L}\times\text{L};\pi^0+\eta+\eta')} = -85(13)$	[132]
$-52(18)$	$-55.60(3)$	$\pi^\pm$ loop, π^0 and resonance exchanges, quark loop	[133]
$-79.2(15.4)$	$-55.60(3)$	$\pi^\pm$ loop, π^0 exchange and quark loop,	[134]
$+83(12)$	$+58(10)$	π^0, η and η' exchanges only	[135]
$+89.6(15.4)$	$+55.60(3)$	$\pi^\pm$ loop, π^0 exchange and quark loop,	[136]
$+83(32)$		ENJL, $a_\mu^{(\text{h. L}\times\text{L};\pi^0+\eta+\eta')} = +85(13)$	[137]

the observation that the pion exchange contribution $a_\mu^{(\text{h. L}\times\text{L};\pi^0)}$ was found to dominate the final values obtained for $a_\mu^{(\text{h. L}\times\text{L})}$, we concentrated our efforts on that part, that I shall now describe in greater detail. For a detailed account on how the other contributions to $a_\mu^{(\text{h. L}\times\text{L})}$ arise, I refer the reader to the original works [131–134].

The contributions to $\Pi_{\mu\nu\lambda\rho}(q_1, q_2, q_3)$ due to single neutral pion exchanges, see Fig. 12, read

$$
\Pi_{\mu\nu\lambda\rho}^{(\pi^0)}(q_1, q_2, q_3) = i\, \frac{\mathcal{F}_{\pi^0\gamma^*\gamma^*}(q_1^2, q_2^2)\, \mathcal{F}_{\pi^0\gamma^*\gamma^*}(q_3^2, (q_1+q_2+q_3)^2)}{(q_1+q_2)^2 - M_\pi^2}
$$
$$
\times \varepsilon_{\mu\nu\alpha\beta}\, q_1^\alpha q_2^\beta\, \varepsilon_{\lambda\rho\sigma\tau}\, q_3^\sigma (q_1+q_2)^\tau
$$
$$
+ i\, \frac{\mathcal{F}_{\pi^0\gamma^*\gamma^*}(q_1^2, (q_1+q_2+q_3)^2)\, \mathcal{F}_{\pi^0\gamma^*\gamma^*}(q_2^2, q_3^2)}{(q_2+q_3)^2 - M_\pi^2}
$$
$$
\times \varepsilon_{\mu\rho\alpha\beta}\, q_1^\alpha (q_2+q_3)^\beta\, \varepsilon_{\nu\lambda\sigma\tau}\, q_2^\sigma q_3^\tau
$$
$$
+ i\, \frac{\mathcal{F}_{\pi^0\gamma^*\gamma^*}(q_1^2, q_3^2)\, \mathcal{F}_{\pi^0\gamma^*\gamma^*}(q_2^2, (q_1+q_2+q_3)^2)}{(q_1+q_3)^2 - M_\pi^2}
$$
$$
\times \varepsilon_{\mu\lambda\alpha\beta}\, q_1^\alpha q_3^\beta\, \varepsilon_{\nu\rho\sigma\tau}\, q_2^\sigma (q_1+q_3)^\tau \ .
$$

$$(5.27)$$

The form factor $\mathcal{F}_{\pi^0\gamma^*\gamma^*}(q_1^2, q_2^2)$, which corresponds to the shaded blobs in Fig. 12, is defined as

$$
i \int d^4x e^{iq\cdot x} \langle \Omega | T\{j_\mu(x) j_\nu(0)\} | \pi^0(p)\rangle = \varepsilon_{\mu\nu\alpha\beta}\, q^\alpha p^\beta\, \mathcal{F}_{\pi^0\gamma^*\gamma^*}(q^2, (p-q)^2),
$$

with $\mathcal{F}_{\pi^0\gamma^*\gamma^*}(q_1^2, q_2^2) = \mathcal{F}_{\pi^0\gamma^*\gamma^*}(q_2^2, q_1^2)$. Inserting the expression (5.27) into (5.25) and computing the corresponding Dirac traces in (5.26), we obtain

$$
a_\mu^{(\text{h. L}\times\text{L};\pi^0)} = e^6 \int \frac{d^4q_1}{(2\pi)^4} \int \frac{d^4q_2}{(2\pi)^4} \frac{1}{[(p+q_1)^2 - m^2][(p-q_2)^2 - m^2]}
$$

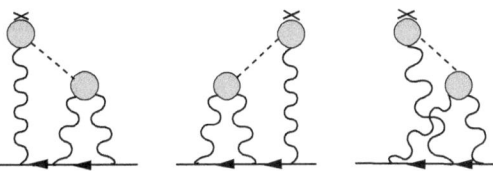

Fig. 12. The pion-pole contributions to light-by-light scattering. The shaded blobs represent the form factor $\mathcal{F}_{\pi^0\gamma^*\gamma^*}$. The first and second graphs give rise to identical contributions, involving the function $T_1(q_1, q_2; p)$ in (5.28), whereas the third graph gives the contribution involving $T_2(q_1, q_2; p)$.

$$
\times \frac{1}{q_1^2 q_2^2 (q_1 + q_2)^2} \left[\frac{\mathcal{F}_{\pi^0\gamma^*\gamma^*}(q_1^2, (q_1 + q_2)^2)\, \mathcal{F}_{\pi^0\gamma^*\gamma^*}(q_2^2, 0)}{q_2^2 - M_\pi^2} \, T_1(q_1, q_2; p) \right.
$$
$$
\left. + \frac{\mathcal{F}_{\pi^0\gamma^*\gamma^*}(q_1^2, q_2^2)\, \mathcal{F}_{\pi^0\gamma^*\gamma^*}((q_1 + q_2)^2, 0)}{(q_1 + q_2)^2 - M_\pi^2} \, T_2(q_1, q_2; p) \right], \qquad (5.28)
$$

where $T_1(q_1, q_2; p)$ and $T_2(q_1, q_2; p)$ denote two polynomials in the invariants $p \cdot q_1$, $p \cdot q_2$, $q_1 \cdot q_2$. Their expressions can be found in [135]. The former arises from the two first diagrams shown in Fig. 12, which give identical contributions, while the latter corresponds to the third diagram on this same figure. At this stage, it should also be pointed out that the expression (5.27) does not, strictly speaking, represent the contribution arising from the pion pole only. The latter would require that the numerators in (5.27) be evaluated at the values of the momenta that correspond to the pole indicated by the corresponding denominators. For instance, the numerator of the term proportional to $T_1(q_1, q_2; p)$ in (5.28) should rather read $\mathcal{F}_{\pi^0\gamma^*\gamma^*}(q_1^2, (q_1^2 + 2q_1 \cdot q_2 + M_\pi^2)\, \mathcal{F}_{\pi^0\gamma^*\gamma^*}(M_\pi^2, 0)$ with $q_2^2 = M_\pi^2$. However, (5.28) corresponds to what previous authors have called the pion pole contribution, and which they had found to dominate the final result.

From here on, information on the form factor $\mathcal{F}_{\pi^0\gamma^*\gamma^*}(q_1^2, q_2^2)$ is required in order to proceed. The simplest model for the form factor follows from the Wess-Zumino-Witten (WZW) term [138,139] that describes the Adler-Bell-Jackiw anomaly [140,141] in chiral perturbation theory. Since in this case the form factor is constant, one needs an ultraviolet cutoff, at least in the contribution to (5.28) involving T_1, the one involving T_2 gives a finite result even for a constant form factor [131]. Therefore, this model cannot be used for a reliable estimate, but at best serves only illustrative purposes in the present context.[21] Previous calculations [131,133,134] have also used the usual vector meson dominance form factor [see also [143]]. The expressions

---

[21] In the context of an effective field theory approach, the pion pole with WZW vertices represents a chirally suppressed, but large-N_C dominant contribution, whereas the charged pion loop is dominant in the chiral expansion, but suppressed in the large-N_C limit [142].

for the form factor $\mathcal{F}_{\pi^0\gamma^*\gamma^*}$ based on the ENJL model that have been used in [132] do not allow a straightforward analytical calculation of the loop integrals. However, compared with the results obtained in [131,133,134], the corresponding numerical estimates are rather close to the VMD case [within the error attributed to the model dependence]. Finally, representations of the form factor $\mathcal{F}_{\pi^0\gamma^*\gamma^*}$, based on the large-$N_C$ approximation to QCD and that takes into account constraints from chiral symmetry at low energies, and from the operator product expansion at short distances, have been discussed in [144] . They involve either one vector resonance [lowest meson dominance, LMD] or two vector resonances [LMD+V], see [144] for details. The four types of form factors just mentioned can be written in the form [F_π is the pion decay constant]

$$\mathcal{F}_{\pi^0\gamma^*\gamma^*}(q_1^2, q_2^2) = \frac{F_\pi}{3} \left[f(q_1^2) - \sum_{M_{V_i}} \frac{1}{q_2^2 - M_{V_i}^2} g_{M_{V_i}}(q_1^2) \right]. \tag{5.29}$$

For the VMD and LMD form factors, the sum in (5.29) reduces to a single term, and the corresponding function is denoted $g_{M_V}(q^2)$. It depends on the mass M_V of the vector resonance, which will be identified with the mass of the ρ meson. For our present purposes, it is enough to consider only these two last cases, along with the constant WZW form factor. The corresponding functions $f(q^2)$ and $g_{M_V}(q^2)$ are displayed in Table 5.

Table 5. The functions $f(q^2)$ and $g_{M_V}(q^2)$ of (5.29) for the different form factors. N_C is the number of colours, taken equal to 3, and $F_\pi = 92.4$ MeV is the pion decay constant. Furthermore, $c_V = \frac{N_C}{4\pi^2} \frac{M_V^4}{F_\pi^2}$.

	$f(q^2)$	$g_{M_V}(q^2)$
WZW	$-\dfrac{N_C}{4\pi^2 F_\pi^2}$	0
VMD	0	$\dfrac{N_C}{4\pi^2 F_\pi^2} \dfrac{M_V^4}{q^2 - M_V^2}$
LMD	$\dfrac{1}{q^2 - M_V^2}$	$-\dfrac{q^2 + M_V^2 - c_V}{q^2 - M_V^2}$

We may now come back to (5.28). With a representation of the form (5.29), the angular integrations can be performed, using standard Gegenbauer polynomial techniques [hyperspherical approach], see [145,146,76]. This leads to a two dimensional integral representation:

$$a_\mu^{(\text{h. L}\times\text{L};\pi^0)} = \left(\frac{\alpha}{\pi}\right)^3 \left[a_\mu^{(\pi^0;1)} + a_\mu^{(\pi^0;2)} \right], \tag{5.30}$$

$$a_\mu^{(\pi^0;1)} = \int_0^\infty dQ_1 \int_0^\infty dQ_2 \left[w_{f_1}(Q_1,Q_2)\, f^{(1)}(Q_1^2,Q_2^2) \right.$$

$$\left. + w_{g_1}(M_V,Q_1,Q_2)\, g_{M_V}^{(1)}(Q_1^2,Q_2^2) \right], \tag{5.31}$$

$$a_\mu^{(\pi^0;2)} = \int_0^\infty dQ_1 \int_0^\infty dQ_2 \sum_{M=M_\pi,M_V} w_{g_2}(M,Q_1,Q_2) g_M^{(2)}(Q_1^2,Q_2^2). \tag{5.32}$$

The functions $f^{(1)}(Q_1^2,Q_2^2)$, $g_{M_V}^{(1)}(Q_1^2,Q_2^2)$, $g_{M_\pi}^{(2)}(Q_1^2,Q_2^2)$ and $g_{M_V}^{(2)}(Q_1^2,Q_2^2)$ are expressed in terms of the functions given in Table 5, see [135], where the universal [for the class of form factors that have a representation of the type shown in (5.29)] weight functions w_{f_1}, w_{g_1}, and w_{g_2} in (5.31) and (5.32) can also be found. The latter are shown in Fig. 13.

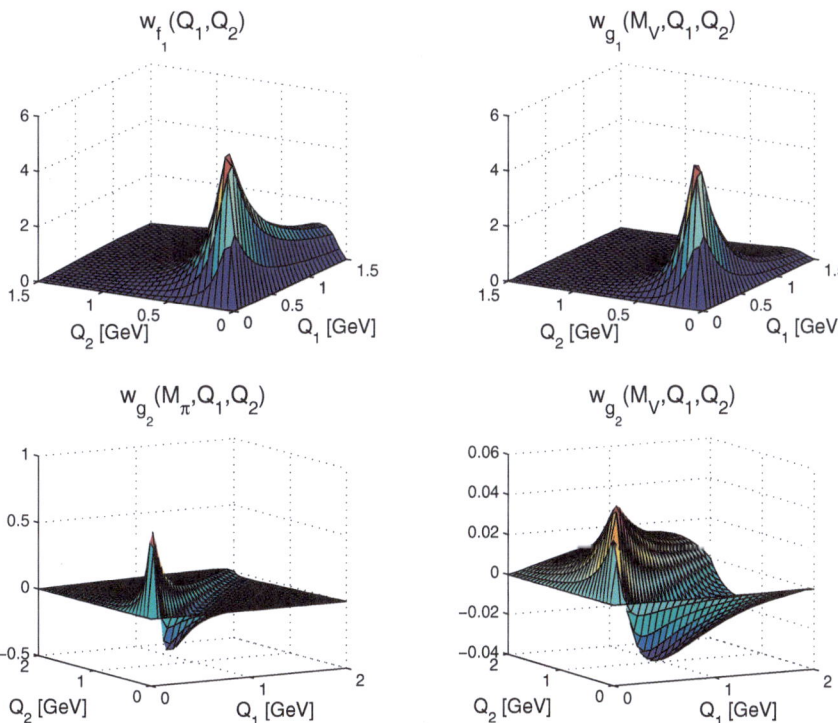

Fig. 13. The weight functions appearing in (5.31) and (5.32). Note the different ranges of Q_i in the subplots. The functions w_{f_1} and w_{g_1} are positive definite and peaked in the region $Q_1 \sim Q_2 \sim 0.5$ GeV. Note, however, the tail in w_{f_1} in the Q_1-direction for $Q_2 \sim 0.2$ GeV. The functions $w_{g_2}(M_\pi,Q_1,Q_2)$ and $w_{g_2}(M_V,Q_1,Q_2)$ take both signs, but their magnitudes remain small as compared to $w_{f_1}(Q_1,Q_2)$ and $w_{g_1}(M_V,Q_1,Q_2)$. We have used $M_V = M_\rho = 770$ MeV.

The functions w_{f_1} and w_{g_1} are positive and concentrated around momenta of the order of 0.5 GeV. This feature was already observed numerically in [132] by varying the upper bound of the integrals [an analogous analysis is contained in [133]]. Note, however, the tail in w_{f_1} in the Q_1 direction for $Q_2 \sim 0.2$ GeV. On the other hand, the function w_{g_2} has positive and negative contributions in that region, which will lead to a strong cancellation in the corresponding integrals, provided they are multiplied by a positive function composed of the form factors [see the numerical results below]. As can be seen from the plots, and checked analytically, the weight functions vanish for small momenta. Therefore, the integrals are infrared finite. The behaviours of the weight functions for large values of Q_1 and/or Q_2 can also be worked out analytically. From these, one can deduce that in the case of the WZW form factor, the corresponding, divergent, integral for $a_\mu^{(\pi^0;1)}$ behaves, as a function of the ultraviolet cut off Λ, as $a_\mu^{(\pi^0;1)} \sim \mathcal{C} \ln^2 \Lambda$, with [135]

$$\mathcal{C} = 3 \left(\frac{N_C}{12\pi} \right)^2 \left(\frac{m_\mu}{F_\pi} \right)^2 = 0.0248 \,. \tag{5.33}$$

The log-squared behaviour follows from the general structure of the integral (5.31) for $a_\mu^{(\pi^0;1)}$ in the case of a constant form factor, as pointed out in [6]. The expression (5.33) of the coefficient $\mathcal{C}$ has been derived independently, in [147], through a renormalization group argument in the effective theory framework.

Table 6. Results for the terms $a_\mu^{(\pi^0;1)}$, $a_\mu^{(\pi^0;2)}$ and for the pion exchange contribution to the anomalous magnetic moment $a_\mu^{(\text{h. L}\times\text{L};\pi^0)}$ according to (5.30) for the different form factors considered. In the WZW model, a cutoff of 1 GeV was used in the first contribution, whereas the second term is ultraviolet finite.

Form factor	$a_\mu^{(\pi^0;1)}$	$a_\mu^{(\pi^0;2)}$	$a_\mu^{(\text{h. L}\times\text{L};\pi^0)} \times 10^{10}$
WZW	0.095	0.0020	12.2
VMD	0.044	0.0013	5.6
LMD	0.057	0.0014	7.3

In the case of the other form factors, the integration over Q_1 and Q_2 is finite and can now be performed numerically. [22] Furthermore, since both the VMD and LMD model tend to the WZW constant form factor as $M_V \to \infty$, the results for $a_\mu^{(\pi^0;1)}$ in these models should scale as $\mathcal{C} \ln^2 M_V^2$ for a large resonance mass. This has been checked numerically, and the value of the coefficient $\mathcal{C}$ obtained that way is in perfect agreement with the value given in

---

[22] In the case of the VMD form factor, an analytical result is now also available [148].

(5.33). The results of the integration over Q_1 and Q_2 are displayed in Table 6. They definitely show a sign difference when compared to those obtained in [131,133,134,143], although in absolute value the numbers agree perfectly. After the results of Table 6 were made public [135], previous authors checked their calculations and, after some time, discovered that they had made a sign mistake at some stage [136,137]. The results presented in Table 6 and in [135,147] have also received independent confirmations [149,148].

The analysis of [135] leads to the following estimates

$$a_\mu^{(\text{h. L}\times\text{L};\pi^0)} = 5.8(1.0) \times 10^{-10}, \qquad (5.34)$$

and

$$a_e^{(\text{h. L}\times\text{L};\pi^0)} = 5.1 \times 10^{-14}. \qquad (5.35)$$

Taking into account the other contributions computed by previous authors, and adopting a conservative attitude towards the error to be ascribed to their model dependences, the total contribution to a_μ coming from the hadronic light-by-light scattering diagrams amounts to

$$a_\mu^{(\text{h. L}\times\text{L})} = 8(4) \times 10^{-10}. \qquad (5.36)$$

As a last remark, let me point out that the contribution depicted in Fig. 9 also involves the four point function $\Pi_{\mu\nu\lambda\rho}(q_1, q_2, q_3)$. It would be interesting to have an evaluation of this contribution to $a_\mu^{(\text{h.v.p. 1})}$ based on the same models that were used to evaluate $a_\mu^{\text{h. L}\times\text{L}}$. The corresponding contribution arising from the neutral pion exchange has been evaluated in [148], but it is not obvious that it also dominates the complete result, since the kinematical configuration is different. This evaluation would also allow a direct comparison with the evaluations of Fig. 9 based on data.

5.3   Electroweak Contributions to a_μ

Electroweak corrections to a_μ have been considered at the one and two loop levels. The one loop contributions, shown in Fig. 14, have been worked out some time ago, and read [150]-[154]

$$a_\mu^{\text{W}(1)} = \frac{G_\text{F}}{\sqrt{2}} \frac{m_\mu^2}{8\pi^2} \left[\frac{5}{3} + \frac{1}{3} \left(1 - 4\sin^2\theta_W\right)^2 + \mathcal{O}\left(\frac{m_\mu^2}{M_Z^2} \log \frac{M_Z^2}{m_\mu^2}\right) \right.$$
$$\left. + \mathcal{O}\left(\frac{m_\mu^2}{M_H^2} \log \frac{M_H^2}{m_\mu^2}\right) \right], \qquad (5.37)$$

where the weak mixing angle is defined by $\sin^2\theta_W = 1 - M_W^2/M_Z^2$. Numerically, with $G_\text{F} = 1.16639(1) \times 10^{-5} \text{ GeV}^{-2}$ and $\sin^2\theta_W = 0.224$,

$$a_\mu^{\text{W}(1)} = 194.8 \times 10^{-11}, \qquad (5.38)$$

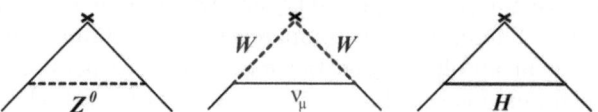

Fig. 14. One loop weak interaction contributions to the anomalous magnetic moment.

It is convenient to separate the two–loop electroweak contributions into two sets of Feynman graphs: those which contain closed fermion loops, which are denoted by $a_\mu^{\text{EW}(2);\text{f}}$, and the others, $a_\mu^{\text{EW}(2);\text{b}}$. In this notation, the electroweak contribution to the muon anomalous magnetic moment is

$$a_\mu^{\text{EW}} = a_\mu^{\text{W}(1)} + a_\mu^{\text{EW}(2);\text{f}} + a_\mu^{\text{EW}(2);\text{b}} . \tag{5.39}$$

I shall review the calculation of the two–loop contributions separately.

Two Loop Bosonic Contributions. The leading logarithmic terms of the two–loop electroweak bosonic corrections have been extracted using asymptotic expansion techniques, see e.g. [155]. In the approximation where $\sin^2 \theta_W \rangle 0$ and $M_H \sim M_W$ these calculations simplify considerably and one obtains

$$a_\mu^{\text{EW}(2);\text{b}} = \frac{G_{\text{F}}}{\sqrt{2}} \frac{m_\mu^2}{8\pi^2} \frac{\alpha}{\pi} \times \left[-\frac{65}{9} \ln \frac{M_W^2}{m_\mu^2} + \mathcal{O}\left(\sin^2 \theta_W \ln \frac{M_W^2}{m_\mu^2} \right) \right] . \tag{5.40}$$

In fact, these contributions have now been evaluated analytically, in a systematic expansion in powers of $\sin^2 \theta_W$, up to $\mathcal{O}[(\sin^2 \theta_W)^3]$, where $\ln \frac{M_W^2}{m_\mu^2}$ terms, $\ln \frac{M_H^2}{M_W^2}$ terms, $\frac{M_W^2}{M_H^2} \ln \frac{M_H^2}{M_W^2}$ terms, $\frac{M_W^2}{M_H^2}$ terms and constant terms are kept [97]. Using $\sin^2 \theta_W = 0.224$ and $M_H = 250\,\text{GeV}$, the authors of [97] find

$$a_\mu^{\text{EW}(2);\text{b}} = \frac{G_{\text{F}}}{\sqrt{2}} \frac{m_\mu^2}{8\pi^2} \frac{\alpha}{\pi} \times \left[-5.96 \ln \frac{M_W^2}{m_\mu^2} + 0.19 \right]$$

$$= \frac{G_{\text{F}}}{\sqrt{2}} \frac{m_\mu^2}{8\pi^2} \left(\frac{\alpha}{\pi} \right) \times [-78.9] = -21.1 \times 10^{-11} , \tag{5.41}$$

showing, in retrospect, that the simple approximation in (5.40) is rather good.

Two Loop Fermionic Contributions. The discussion of the two–loop electroweak fermionic corrections is more delicate. First, it contains a hadronic contribution. Next, because of the cancellation between lepton loops and quark loops in the electroweak $U(1)$ anomaly, one cannot separate hadronic effects from leptonic effects any longer. In fact, as discussed in [156,157], it is this cancellation which eliminates some of the large logarithms which, incorrectly were kept in [158]. It is therefore appropriate to separate the two–loop electroweak fermionic corrections into two classes: One is the class arising from Feynman diagrams containing a lepton or a quark loop, with the external photon, a virtual photon and a virtual Z^0 attached to it, see Fig. 15.[23] The quark loop of course again represents non perturbative hadronic contributions which have to be evaluated using some model. This first class is denoted by $a_\mu^{\mathrm{EW}(2);\mathrm{f}}(\ell; q)$. It involves the QCD correlation function

$$W_{\mu\nu\rho}(q, k) = \int d^4x\, e^{iq\cdot x} \int d^4y\, e^{i(k-q)\cdot y} \langle \Omega\, |\mathrm{T}\{j_\mu(x) A_\nu^{(Z)}(y) j_\rho(0)\}|\Omega \rangle\,,\quad (5.42)$$

with k the incoming external photon four-momentum associated with the classical external magnetic field. As previously, j_ρ denotes the hadronic part of the electromagnetic current, and $A_\rho^{(Z)}$ is the axial component of the current which couples the quarks to the Z^0 gauge boson. The other class is defined by the rest of the diagrams, where quark loops and lepton loops can be treated separately, and is called $a_\mu^{\mathrm{EW}(2);\mathrm{f}}(\text{residual})$.

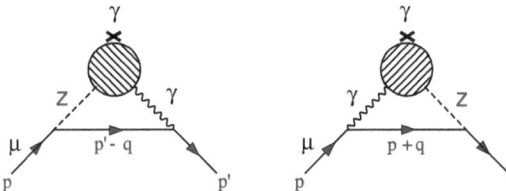

Fig. 15. Graphs with hadronic contributions to $a_\mu^{\mathrm{EW}(2);\mathrm{f}}(\ell, q)$ and involving the QCD three point function $W_{\mu\nu\rho}(q, k)$.

The contribution from $a_\mu^{\mathrm{EW}(2);\mathrm{f}}(\text{residual})$ brings in factors of the ratio m_t^2/M_W^2. It has been evaluated, to a very good approximation, in [157], with the result

$$a_\mu^{\mathrm{EW}(2);\mathrm{f}}(\text{residual}) = \frac{G_\mathrm{F}}{\sqrt{2}} \frac{m_\mu^2}{8\pi^2} \frac{\alpha}{\pi} \times \left[\frac{1}{2\sin^2\theta_W} \left(-\frac{5}{8} \frac{m_t^2}{M_W^2} - \log\frac{m_t^2}{M_W^2} - \frac{7}{3} \right) \right.$$
$$\left. + \Delta_{\mathrm{Higgs}} \right]\,,\qquad (5.43)$$

where Δ_{Higgs} denotes the contribution from diagrams with Higgs lines, which the authors of [157] estimate to be $\Delta_{\mathrm{Higgs}} = -5.5 \pm 3.7$, and therefore,

$$a_\mu^{\mathrm{EW}(2);\mathrm{f}}(\text{residual}) = \frac{G_\mathrm{F}}{\sqrt{2}} \frac{m_\mu^2}{8\pi^2} \frac{\alpha}{\pi} \times [-21(4)] = -5.6(1.4) \times 10^{-11}\,.\quad (5.44)$$

---

[23] If one works in a renormalizable gauge, the contributions where the Z^0 is replaced by the neutral unphysical Higgs should also be included. The final result does not depend on the gauge fixing parameter ξ_Z, if one works in the class of 't Hooft gauges.

Let us finally discuss the contributions to $a_\mu^{\text{EW}(2);\text{f}}(\ell; q)$. Here, it is convenient to treat the contributions from the three generations separately. The contribution from the third generation can be calculated in a straightforward way, with the result [156,157]

$$a_\mu^{\text{EW}(2);\text{f}}(\tau; t, b) = \frac{G_{\text{F}}}{\sqrt{2}} \frac{m_\mu^2}{8\pi^2} \frac{\alpha}{\pi} \times \left[-3\ln\frac{M_Z^2}{m_\tau^2} - \ln\frac{M_Z^2}{m_b^2} - \frac{8}{3}\ln\frac{m_t^2}{M_Z^2} + \frac{8}{3} \right. $$
$$\left. + \mathcal{O}\left(\frac{M_Z^2}{m_t^2}\ln\frac{m_t^2}{M_Z^2}\right) \right]$$

(5.45)

$$= \frac{G_{\text{F}}}{\sqrt{2}} \frac{m_\mu^2}{8\pi^2} \frac{\alpha}{\pi} \times [-30.6] = -8.2 \times 10^{-11}. \tag{5.46}$$

In fact the terms of $\mathcal{O}\left(\frac{M_Z^2}{m_t^2}\ln\frac{m_t^2}{M_Z^2}\right)$ and $\mathcal{O}\left(\frac{M_Z^2}{m_t^2}\right)$ have also been calculated in [157]. There are in principle QCD perturbative corrections to this estimate, which have not been calculated, but the result in (5.45) is good enough for the accuracy required at present. The contributions of the remaining charged standard model fermions involve the light quarks u and d, as well as the second generation s quark, for which non perturbative effects tied to the spontaneous breaking of chiral symmetry are important [156,159]. The contributions from the first and second generation are thus most conveniently taken together, with the result

$$a_\mu^{\text{EW}(2);\text{f}}(e, \mu; u, d, s, c) = \frac{G_{\text{F}}}{\sqrt{2}} \frac{m_\mu^2}{8\pi^2} \frac{\alpha}{\pi} \times \left\{ -3\ln\frac{M_Z^2}{m_\mu^2} - \frac{5}{2} \right.$$

(5.47)

$$-3\ln\frac{M_Z^2}{m_\mu^2} + 4\ln\frac{M_Z^2}{m_c^2} - \frac{11}{6} + \frac{8}{9}\pi^2 - 8$$

(5.48)

$$\left. + \left[\frac{4}{3}\ln\frac{M_Z^2}{m_\mu^2} + \frac{2}{3} + \mathcal{O}\left(\frac{m_\mu^2}{M_Z^2}\ln\frac{M_Z^2}{m_\mu^2}\right) \right] + 4.57(1.80) + 0.04(2) \right\}$$

(5.49)

$$= \frac{G_{\text{F}}}{\sqrt{2}} \frac{m_\mu^2}{8\pi^2} \frac{\alpha}{\pi} \times [-28.5(1.8)] = -7.6(5) \times 10^{-11}, \tag{5.50}$$

where the first line shows the result from the e loop and the second line the result from the μ loop and the c quark, which is treated as a heavy quark. The term between brackets in the third line is the one induced by the anomalous term in the hadronic three point function $W_{\mu\nu\rho}(q, k)$ The other contributions have been estimated on the basis of an approximation to the large-N_C limit of QCD, similar to the one discussed for the two-point function $\Pi(Q^2)$ after (5.15), see [159] for details.

The result in (5.47) for the contribution from the first and second generations of quarks and leptons is conceptually rather different from the corresponding one proposed in [157] ,

$$a_\mu^{\mathrm{EW}(2);\mathrm{f}}(e,\mu;u,d,s,c) = \frac{G_\mathrm{F}}{\sqrt{2}} \frac{m_\mu^2}{8\pi^2} \frac{\alpha}{\pi} \left[-3\ln\frac{M_Z^2}{m_\mu^2} + 4\ln\frac{M_Z^2}{m_u^2} - \ln\frac{M_Z^2}{m_d^2} - \frac{5}{2} - 6 \right.$$

$$\tag{5.51}$$

$$\left. -3\ln\frac{M_Z^2}{m_\mu^2} + 4\ln\frac{M_Z^2}{m_c^2} - \ln\frac{M_Z^2}{m_s^2} - \frac{11}{6} + \frac{8}{9}\pi^2 - 6 \right]$$

$$\tag{5.52}$$

$$= \frac{G_\mathrm{F}}{\sqrt{2}} \frac{m_\mu^2}{8\pi^2} \frac{\alpha}{\pi} \times [-31.9] = -8.5 \times 10^{-11} , \tag{5.53}$$

where the light quarks were, *arbitrarily*, treated the same way as heavy quarks, with $m_u = m_d = 0.3\,\mathrm{GeV}$, and $m_s = 0.5\,\mathrm{GeV}$. Numerically, the two expression lead to similar results, though. A more recent analysis [160] provides a non perturbative treatment of the light quark sector, and gives

$$a_\mu^{\mathrm{EW}(2);\mathrm{f}}(e,\mu;u,d,s,c) = \frac{G_\mathrm{F}}{\sqrt{2}} \frac{m_\mu^2}{8\pi^2} \frac{\alpha}{\pi} \times [-24.6] = -6.6 \times 10^{-11} . \tag{5.54}$$

The difference between the two results in (5.47) and (5.54), which is numerically very small, is connected to interesting issues involving the anomalous $\langle VVA \rangle$ three point function in QCD.[24]

The authors of [160] have also performed a detailed renormalization group analysis of the leading logarithm contributions at three loops [25], and found them to be negligible. Taking into account other small effects that were previously neglected, their final value reads [160]

$$a_\mu^{\mathrm{EW}} = 15.4(3) \times 10^{-10} , \tag{5.55}$$

which shows that the two–loop correction represents indeed a reduction of the one–loop result by an amount of 23%. The final error here includes uncertainties in the hadronic part, the variation of the Higgs mass in a range $114\ \mathrm{GeV} \leq M_H \leq 250\ \mathrm{GeV}$, the uncertainty on the mass of the top quark, and unknown three loop effects.

5.4   Comparison with Experiment

We may now put all the pieces together and obtain the value for a_μ predicted by the standard model. We have seen that in the case of the hadronic vacuum

---

[24] Actually, the discussion centers around the transverse, i.e. *non anomalous* part of this QCD correlator, and is related to the existence of *non renormalization theorems* for it [161,160,162].

[25] See also [163].

polarization contributions, the latest evaluation [122] shows a discrepancy between the value obtained exclusively from e^+e^- data and the value that arises if τ data are also included. This gives us the two possibilities

$$a_\mu^{SM}(e^+e^-) = (11\,659\,167.5 \pm 7.5 \pm 4.0 \pm 0.4) \times 10^{-10}$$
$$a_\mu^{SM}(\tau) = (11\,659\,192.7 \pm 5.9 \pm 4.0 \pm 0.4) \times 10^{-10}, \qquad (5.56)$$

where the first error comes from hadronic vacuum polarization, the second from hadronic light-by-light scattering, and the last from the QED and weak corrections. When compared to the present experimental average

$$a_\mu^{exp} = (11\,659\,203 \pm 8) \times 10^{-10} \qquad (5.57)$$

there results a difference,

$$a_\mu^{exp} - a_\mu^{SM}(e^+e^-) = 35.5(11.7) \times 10^{-10},$$
$$a_\mu^{exp} - a_\mu^{SM}(\tau) = 10.3(10.7) \times 10^{-10}, \qquad (5.58)$$

which represents 3.0 and 1.0 standard deviations, respectively.

Although experiment and theory have now both reached the same level of accuracy, $\sim \pm 8 \times 10^{-10}$ or 0.7 ppm, the present discrepancy between the e^+e^- and τ based evaluations makes the interpretation of the above results a delicate issue as far as evidence for new physics is concerned. Other evaluations of comparable accuracy [119,120,58] cover a similar range of variation in the difference between experiment and theory. Furthermore, the value obtained for $a_\mu^{SM}(e^+e^-)$ relies strongly on the low-energy data obtained by the CMD-2 experiment, with none of the older data able to check them at the same level of precision. There seems to be an error in these data from the CMD-2 experiment [8], but this clearly needs to be confirmed. In this respect, the prospects for additional high statistics data in the future, either from KLOE or from BaBar, are most welcome. On the other hand, if the present discrepancy in the evaluations of the hadronic vacuum polarization finds a solution in the future, and if the experimental error is further reduced, by, say, a factor of two, then the theoretical uncertainty on the hadronic light-by-light scattering will constitute the next serious limitation on the theoretical side. It is certainly worthwhile to devote further efforts to a better understanding of this contribution, for instance by finding ways to feed more constraints with a direct link to QCD into the descriptions of the four-point function $\Pi_{\mu\nu\rho\sigma}(q_1, q_2, q_3)$.

6   Concluding Remarks

With this review, I hope to have convinced the reader that the subject of the anomalous magnetic moments of the electron and of the muon is an exciting and fascinating topic. It provides a good example of mutual stimulation and strong interplay between experiment and theory.

The anomalous magnetic moment of the electron constitutes a very stringent test of QED and of the practical working of the framework of perturbatively renormalized quantum field theory at higher orders. It tests the validity of QED at very short distances, and provides at present the best determination of the fine structure constant.

The anomalous magnetic moment of the muon represents the best compromise between sensitivity to new degrees of freedom describing physics beyond the standard model and experimental feasibility. Important progress has been achieved on the experimental side during the last couple of years, with the results of the E821 collaboration at BNL. The experimental value of a_μ is now known with an accuracy of 0.7ppm. Hopefully, the Brookhaven experiment will be given the opportunity to reach its initial goal of achieving a measurement at the 0.35 ppm level.

As can be inferred from the examples mentioned in this text, the subject constitutes, from a theoretical point of view, a difficult and error prone topic, due to the technical difficulties encountered in the higher loop calculations. The theoretical predictions have reached a precision comparable to the experimental one, but unfortunately there appears a discrepancy between the most recent evaluations of the hadronic vacuum polarization according to whether τ data are considered or not. Hopefully, this situation will be clarified soon. Hadronic contributions, especially from vacuum polarization and from light-by-light scattering, are responsible for the bulk part of the final uncertainty in the theoretical value a_μ^{SM}. Further efforts are needed in order to bring these aspects under better control.

Acknowledgements

I wish to thank A. Nyffeler, S. Peris, M. Perrottet, and E. de Rafael for stimulating and very pleasant collaborations. Most of the figures appearing in this text were kindly provided by M. Perrottet. A countless number of very informative and useful comments was provided by A. Nyffeler, M. Perrottet, and E. de Rafael. Finally, I wish to thank the organizers of the 41th edition of the Schladming school for their invitation to present these lectures and for providing, together with the students and the other lecturers, a very pleasant and fruitful atmosphere. This work is supported in part by the EC contract No. HPRN-CT-2002-00311 (EURIDICE).

Note Added: Only a few weeks after the lines above had been written, important changes were reported by the CMD-2 collaboration. Several errors were discovered in the analysis underlying the previously published results of [110], and a corrected version is now available [164]. The effects of these changes on the anomalous magnetic moment of the muon have been evaluated in [165], from which we quote

$$a_\mu^{(\text{h.v.p. 1})}(e^+e^-) = 6963(72) \times 10^{-11} \,,$$

and which has to be compared to the average (5.13) based on the previous CMD-2 data. Incorporating a new experimental result [166] for the branching ratio of the $\tau \to \nu_\tau h^- \pi^0$ mode, [165] also gives an updated value

$$a_\mu^{(\text{h.v.p. 1})}(\tau) = 7110(61) \times 10^{-11} \,.$$

Here, the change as compared to the penultimate entry in Table 3 is much smaller. As a result, the two determinations are now almost compatible. With these changes included, one then obtains

$$a_\mu^{\text{SM}}(e^+e^-) = (11\,659\,180.0 \pm 7.2 \pm 4.0 \pm 0.4) \times 10^{-10}$$
$$a_\mu^{\text{SM}}(\tau) = (11\,659\,194.7 \pm 5.9 \pm 4.0 \pm 0.4) \times 10^{-10} \,.$$

When compared to the experimental average (5.57), the resulting differences,

$$a_\mu^{\text{exp}} - a_\mu^{\text{SM}}(e^+e^-) = 22.6(11.3) \times 10^{-10} \,,$$
$$a_\mu^{\text{exp}} - a_\mu^{\text{SM}}(\tau) = 8.3(10.5) \times 10^{-10} \,,$$

represent 2.0 and 0.8 standard deviations, respectively.

References

1. T. Kinoshita Ed., *Quantum Electrodynamics*, World Scientific Publishing Co. Pte. Ltd., 1990.
2. B. E. Lautrup, A. Peterman and E. de Rafael, Phys. Rept. **3**, 193 (1972).
3. J. Calmet, S. Narison, M. Perrottet and E. de Rafael, Rev. Mod. Phys. **49**, 21 (1977).
4. A. Czarnecki and W. J. Marciano, Nucl. Phys. B (Proc. Suppl.) **76**, 245 (1998).
5. V. W. Hughes and T. Kinoshita, Rev. Mod. Phys. **71**, S133 (1999).
6. K. Melnikov, Int. J. Mod. Phys. A **16**, 4591 (2001).
7. E. de Rafael, arXiv:hep-ph/0208251.
8. A. Nyffeler, to appear in the proceedings of the *38th Rencontres de Moriond on Electroweak Interactions and Unified Theories*, Les Arcs, 15-22 March 2003, and arXiv:hep-ph/0305135.
9. A. Czarnecki and W. J. Marciano, Phys. Rev. D **64**, 013014 (2001).
10. A list of recent papers on the subject can be found under the URL http://www.slac.stanford.edu/spires/find/hep/www?c=PRLTA,86,2227.
11. L. L. Foldy, Phys. Rev. **87**, 688 (1952); Rev. Mod. Phys. **30**, 471 (1958).
12. Ya. B. Zeldovich, Soviet Phys. JETP **6**, 1184 (1958).
13. Ya. B. Zeldovich and A. M. Perelomov, Soviet Phys. JETP **12**, 777 (1961).

14. R. E. Marshak, Riazuddin, and C. P. Ryan, *Theory of Weak Interactions in Particle Physics*, John Wiley and Sons Inc., 1969.
15. S. J. Brodsky and J. D. Sullivan, Phys. Rev. **156**, 1644 (1967).
16. R. Barbieri, J. A. Mignaco and E. Remiddi, Nuovo Cimento **11A**, 824 (1972).
17. H. Pietschmann, Zeit. f. Phys. **178**, 409 (1964).
18. T. Appelquist and J. Carazzone, Phys. Rev. D **11**, 2856 (1975).
19. C. Bouchiat, J. Iliopoulos and P. Meyer, Phys. Lett. **B38**, 519 (1972).
20. D. J. Gross and R. Jackiw, Phys. Rev. D **6**, 477 (1972).
21. C. P. Korthals Altes and M. Perrottet, Phys. Lett. B **39**, 546 (1972).
22. T. Sterling and M. J. Veltman, Nucl. Phys. **B 189**, 557 (1981).
23. E. d'Hoker and E. Farhi, Nucl. Phys. **B248**, 59, 77 (1984).
24. T. Kinoshita, Nuovo Cimento **51B**, 140 (1967).
25. B. E. Lautrup and E. de Rafael, Nucl. Phys. **B70**, 317 (1974).
26. E. de Rafael and J. L. Rosner, Ann. Phys. (N. Y.) **82**, 369 (1974).
27. T. Kinoshita and W. J. Marciano, *Theory of the Muon Anomalous Magnetic Moment*, in [1], p. 419.
28. B. Kayser, Phys. Rev. D **26**, 1662 (1982).
29. R. N. Mohapatra and P. B. Pal, *Massive Neutrinos in Physics and Astrophysics*, World Scientific Publishing Co. Pte. Ltd., 1991.
30. J. E. Nafe, E. B. Nelson and I. I. Rabi, Phys. Rev. **71**, 914 (1947).
31. H. G. Dehmelt, Phys. Rev. **109**, 381 (1958).
32. R. S. Van Dyck, P. B. Schwinberg and H. G. Dehmelt, Phys. Rev. Lett. **59**, 26 (1987).
33. P. J. Mohr and B. N. Taylor, Rev. Mod. Phys. **72**, 351 (2000).
34. K. Hagiwara *et al.* [Particle Data Group Collaboration], Phys. Rev. D **66**, 010001 (2002).
35. R. S. Van Dyck, *Anomalous Magnetic Moment of Single Electrons and Positrons: Experiment*, in [1], p. 322.
36. A. Rich and J. C. Wesley, Rev. Mod. Phys. **44**, 250 (1972).
37. P. Kusch and H. M. Fowley, Phys. Rev. **72**, 1256 (1947).
38. P. A. Franken and S. Liebes Jr., Phys. Rev. **104**, 1197 (1956).
39. A. A. Schuppe, R. W. Pidd, and H. R. Crane, Phys. Rev. **121**, 1 (1961).
40. D. T. Wilkinson and H. R. Crane, Phys. Rev. **130**, 852 (1963).
41. G. Gräff, E. Klempt and G. Werth, Z. Phys. **222**, 201 (1968).
42. J. C. Wesley and A. Rich, Phys. Rev. A **4**, 1341 (1971).
43. R. S. Van Dyck, P. B. Schwinberg and H. G. Dehmelt, Phys. Rev. Lett. **38**, 310 (1977).
44. F. J. M. Farley and E. Picasso, *The Muon g - 2 Experiments*, in [1], p. 479.
45. J. Bailey et al., Phys. Lett. **B28**, 287 (1968).
46. J. Bailey et al., Phys. Lett. **B55**, 420 (1975).
47. J. Bailey et al. [CERN-Mainz-Daresbury Collaboration], Nucl. Phys. **B 150**, 1 (1979).
48. R. M. Carey et al.,Phys. Rev. Lett. **82**, 1632 (1999).
49. H. N. Brown et al.[Muon (g - 2) Collaboration], Phys. Rev. D **62**, 091101(R) (2000).
50. H. N. Brown et al.[Muon (g - 2) Collaboration], Phys. Rev. Lett. **86**, 2227 (2001).
51. G. W. Bennett et al.[Muon (g - 2) Collaboration], Phys. Rev. Lett. **89**, 101804 (2002); Erratum-ibid. **89**, 129903 (2002).

52. E. D. Commins, S. B. Ross, D. DeMille and B. C. Regan, Phys. Rev. A **50**, 2960 (1994).

53. B. C. Regan, E. D. Commins, C. J. Schmidt and D. DeMille, Phys. Rev. Lett. **88**, 071805 (2002).

54. J. L. Feng, K. T. Matchev and Y. Shadmi, Nucl. Phys. **B 613**, 366 (2001); Phys. Lett. **B555**, 89 (2003).

55. Y. K. Semertzidis *et al.*, Int. J. Mod. Phys. A **16S1B**, 690 (2001).

56. K. Ackerstaff et al. [OPAL Collaboration], Phys. Lett. **B431**, 188 (1998).

57. M. Acciarri et al. [L3 Collaboration], Phys. Lett. **B434**, 169 (1998).

58. S. Narison, Phys. Lett. **B513**, 53 (2001); Erratum-ibid. **B526**, 414 (2002).

59. J. Schwinger, Phys. Rev. **73**, 413 (1948); **76**, 790 (1949).

60. R. Karplus and N. M. Kroll, Phys. Rev. **77**, 536 (1950).

61. A. Peterman, Helv. Phys. Acta **30**, 407 (1957).

62. C. M. Sommerfield, Phys. Rev. **107**, 328 (1957).

63. C. M. Sommerfield, Ann. Phys. (N.Y.) **5**, 26 (1958).

64. G. S. Adkins, Phys. Rev. D **39**, 3798 (1989).

65. J. Schwinger, *Particles, Sources and Fields, VolumeIII*, Addison-Wesley Publishing Company, Inc., 1989.

66. D. Kreimer, arXiv:hep-th/9412045; D. Kreimer, J. Knot Theor. Ramifications **6**, 479 (1997); D. J. Broadhurst, J. A. Gracey and D. Kreimer, Z. Phys. C **75**, 559 (1997); D. J. Broadhurst and D. Kreimer, Phys. Lett. **B393**, 403 (1997).

67. D. Kreimer, Adv. Theor. Math. Phys. **2**, 303 (1998).

68. A. Connes and D. Kreimer, Commun. Math. Phys. **199**, 203 (1998).

69. S. Laporta and E. Remiddi, Phys. Lett. **B379**, 283 (1996).

70. R. Barbieri and E. Remiddi, Nucl. Phys. **B 90**, 233 (1975).

71. M. A. Samuel and G. Li, Phys. Rev. D **44**, 3935 (1991); ibid. D **46**, 4782 (1992) and D **48**, 1879 (1993), errata.

72. S. Laporta and E. Remiddi, Phys. Lett. **B265**, 181 (1991).

73. S. Laporta, Phys. Rev. D **47**, 4793 (1993).

74. S. Laporta, Phys. Lett. **B343**, 421 (1995).

75. S. Laporta and E. Remiddi, Phys. Lett. **B356**, 390 (1995).

76. R. Z. Roskies, E. Remiddi and M. J. Levine, *Analytic evaluation of sixth-order contributions to the electron's g factor*, in [1], p. 162.

77. T. Kinoshita, *Theory of the anomalous magnetic moment of the electron – Numerical Approach*, in [1], p. 218.

78. J. Aldins, S. J. Brodsky, A. Dufner, and T. Kinoshita, Phys. Rev. Lett. **23**, 441 (1970); Phys. Rev. D **1**, 2378 (1970).

79. S. J. Brodsky and T. Kinoshita, Phys. Rev. D **3**, 356 (1971).

80. J. Calmet and M. Perrottet, Phys. Rev. D **3**, 3101 (1971).

81. J. Calmet and A. Peterman, Phys. Lett. **B47**, 369 (1973).

82. M. J. Levine and J. Wright, Phys. Rev. Lett. **26**, 1351 (1971); Phys. Rev. D **8**, 3171 (1973).

83. R. Carroll and Y. P. Yao, Phys. Lett. **B48**, 125 (1974).

84. P. Cvitanovic and T. Kinoshita, Phys. Rev. D **10**, 3978, 3991, 4007 (1974).

85. T. Kinoshita and W. B. Lindquist, Phys. Rev. D **27**, 867, 877, 886 (1983); D **39**, 2407 (1989); D **42**, 636 (1990).

86. M. Caffo, S. Turrini, and E. Remiddi, Phys. Rev. D **30**, 483 (1984).

87. E. Remiddi and S. P. Sorella, Lett. Nuovo Cim. **44**, 231 (1985).

88. T. Kinoshita, IEEE Trans. Instrum. Meas. **44**, 498 (1995).

89. T. Kinoshita and M. Nio, Phys. Rev. Lett. **90**, 021803 (2003).
90. H. Suura and E. Wichmann, Phys. Rev. **105**, 1930 (1957).
91. A. Peterman, Phys. Rev. **105**, 1931 (1957).
92. H. H. Elend, Phys. Lett. **20**, 682 (1966); Erratum-ibid. **21**, 720 (1966).
93. B. E. Lautrup and E. de Rafael, Phys. Rev. **174**, 1835 (1968).
94. W. Liu *et al.*, Phys. Rev. Lett. **82**, 711 (1999).
95. S. Laporta, Nuovo Cimento **106A**, 675 (1993).
96. S. Laporta and E. Remiddi, Phys. Lett. **B301**, 440 (1993).
97. A. Czarnecki, B. Krause and W. J. Marciano, Phys. Rev. Lett. **76**, 3267 (1996).
98. B. Krause, Phys. Lett. **B390**, 392 (1997).
99. B. E. Lautrup, Phys. Lett. **B32**, 627 (1970).
100. B. E. Lautrup and E. de Rafael, Nuovo Cim. **64A**, 322 (1970).
101. B. E. Lautrup, A. Peterman and E. de Rafael, Nuovo Cim. **1A**, 238 (1971).
102. T. Kinoshita, Phys. Rev. D **47**, 5013 (1993).
103. T. Kinoshita, B. Nizic, Y. Okamoto, Phys. Rev. D **41**, 593 (1990).
104. A. S. Yelkhovsky, Sov. J. Nucl. Phys. **49**, 656 (1989).
105. A. I. Milstein and A. S. Yelkhovsky, Phys. Lett. **B233**, 11 (1989).
106. S. G. Karshenboim, Phys. Atom. Nucl. **56**, 857 (1993).
107. C. Bouchiat and L. Michel, J. Phys. Radium **22**, 121 (1961).
108. L. Durand III, Phys. Rev. **128**, 441 (1962); Erratum-ibid. **129**, 2835 (1963).
109. J. Z. Bai et al. [BES Collaboration], Phys. Rev. Lett. **84**, 594 (2000); Phys. Rev. Lett. **88**, 101802 (2000).
110. R. R. Akhmetshin et al. [CMD-2 Collaboration], Phys. Lett. **B527**, 161 (2002).
111. R. Barate et al. [ALEPH Collaboration], Z. Phys. **C 2**, 123 (1997).
112. K. Ackerstaff at al. [OPAL Collaboration], Eur. J. Phys. C **7**, 571 (1999).
113. S. Anderson et al. [CLEO Collaboration], Phys. Rev. D **61**, 112002 (2000).
114. K. W. Edwards et al. [CLEO Collaboration], Phys. Rev. D **61**, 072003 (2000).
115. S. Eidelman and F.Jegerlehner, Z. Phys. C **67**, 585 (1995).
116. D. H. Brown and W. A. Worstell, Phys. Rev. D **54**, 3237 (1996).
117. R. Alemany, M. Davier and A. Höcker, Eur. Phys. J. **C 2**, 123 (1998).
118. M. Davier and A. Höcker, Phys. Lett. **B419**, 419 (1998).
119. M. Davier and A. Höcker, Phys. Lett. **B435**, 427 (1998).
120. J. F. De Trocóniz and F. J. Ynduráin, Phys. Rev. D **65**, 093001 (2002).
121. F. Jegerlehner, J. Phys. G **29**, 101 (2003).
122. M. Davier, S. Eidelman, A. Höcker and Z. Zhang, Eur. Phys. J. C **27**, 497 (2003).
123. K. Hagiwara, A. D. Martin, D. Nomura and T. Teubner, Phys. Lett. B **557**, 69 (2003).
124. A. G. Denig *et al.* [the KLOE Collaboration], arXiv:hep-ex/0211024.
125. E. P. Solodov [BABAR collaboration], in *Proc. of the e^+e^- Physics at Intermediate Energies Conference* ed. Diego Bettoni, and arXiv:hep-ex/0107027.
126. M. Perrottet and E. de Rafael, unpublished.
127. S. Peris, M. Perrottet and E. de Rafael, JHEP **9805**, 011 (1998).
128. G. 't Hooft, Nucl. Phys. **B 72**, 461 (1974).
129. E. Witten, Nucl. Phys. **B 160**, 157 (1979).
130. J. Calmet, S. Narison, M. Perrottet and E. de Rafael, Phys. Lett. **B61**, 283 (1976).
131. T. Kinoshita, B. Nizic, Y. Okamoto, Phys. Rev. D **31**, 2108 (1985).
132. J. Bijnens, E. Pallante and J. Prades, Nucl. Phys. B **474**, 379 (1996).

133. M. Hayakawa, T. Kinoshita and A. I. Sanda, Phys. Rev. Lett. **75**, 790 (1995); Phys. Rev. D **54**, 3137 (1996).

134. M. Hayakawa and T. Kinoshita, Phys. Rev. D **57**, 465 (1998).

135. M. Knecht and A. Nyffeler, Phys. Rev. D **65**, 073034 (2002).

136. M. Hayakawa and T. Kinoshita, arXiv:hep-ph/0112102, and the erratum to [134] published in Phys. Rev. D **66**, 019902(E) (2002).

137. J. Bijnens, E. Pallante and J. Prades, Nucl. Phys. **B 626**, 410 (2002).

138. J. Wess and B. Zumino, Phys. Lett. **37B**, 95 (1971).

139. E. Witten, Nucl. Phys. **B223**, 422 (1983).

140. S. L. Adler, Phys. Rev. **177**, 2426 (1969).

141. J. S. Bell and R. Jackiw, Nuovo Cimento A **60**, 47 (1969).

142. E. de Rafael, Phys. Lett. **B322**, 239 (1994).

143. J. Bijnens and F. Persson, arXiv:hep-ph/0106130.

144. M. Knecht and A. Nyffeler, Eur. Phys. J. C **21**, 659 (2001).

145. J. L. Rosner, Ann. Phys. (N.Y.) **44**, 11 (1967).

146. M. J. Levine and R. Roskies, Phys. Rev. D **9**, 421 (1974); M. J. Levine, E. Remiddi, and R. Roskies, *ibid.* **20**, 2068 (1979).

147. M. Knecht, A. Nyffeler, M. Perrottet and E. de Rafael, Phys. Rev. Lett. **88**, 071802 (2002).

148. I. Blokland, A. Czarnecki and K. Melnikov, Phys. Rev. Lett. **88**, 071803 (2002).

149. W. J. Bardeen and A. de Gouvea, private communication.

150. W.A. Bardeen, R. Gastmans and B.E. Lautrup, Nucl. Phys. **B46**, 315 (1972).

151. G. Altarelli, N. Cabbibo and L. Maiani, Phys. Lett. **40B**, 415 (1972).

152. R. Jackiw and S. Weinberg, Phys. Rev. D **5**, 2473 (1972).

153. I. Bars and M. Yoshimura, Phys. Rev. D **6**, 374 (1972).

154. M. Fujikawa, B.W. Lee and A.I. Sanda, Phys. Rev. D **6**, 2923 (1972).

155. V.A. Smirnov, Mod. Phys. Lett. A **10**, 1485 (1995).

156. S. Peris, M. Perrottet and E. de Rafael, Phys. Lett. **B355**, 523 (1995).

157. A. Czarnecki, B. Krause and W. Marciano, Phys. Rev. D **52**, R2619 (1995).

158. T.V. Kukhto, E.A. Kuraev, A. Schiller and Z.K. Silagadze, Nucl. Phys. **B 371**, 567 (1992).

159. M. Knecht, S. Peris, M. Perrottet and E. de Rafael, JHEP **0211**, 003 (2002).

160. A. Czarnecki, W. J. Marciano and A. Vainshtein, Phys. Rev. D **67**, 073006 (2003).

161. A. Vainshtein, arXiv:hep-ph/0212231.

162. M. Knecht, S. Peris, M. Perrottet and E. de Rafael, in preparation.

163. G. Degrassi and G.F. Giudice, Phys. Rev. **58D** (1998) 053007.

164. R. R. Akhmetshin et al. [CMD-2 Collaboration], arXiv:hep-ex/0308008.

165. M. Davier, S. Eidelman, A. Höcker and Z. Zhang, arXiv:hep-ph/0308213.

166. P. Achard et al. [L3 Collaboration], CERN-EP/2003-19.

CP Violation in B and K Decays: 2003

Andrzej J. Buras

Technical University Munich, Physics Department, 85748 Garching, Germany,
Andrzej_Buras@ph.tum.de

1   Introduction

1.1   Preface

CP violation in B and K meson decays is not surprisingly one of the central topics in particle physics. Indeed, CP-violating and rare decays of K and B mesons are very sensitive to the flavour structure of the Standard Model (SM) and its extensions. In this context a very important role is played by the Cabibbo-Kobayashi-Maskawa (CKM) matrix [1,2] that parametrizes the weak charged current interactions of quarks: its departure from the unit matrix is the origin of all flavour violating and CP-violating transitions in the SM and its simplest extensions.

One of the important questions still to be answered, is whether the CKM matrix is capable to describe with its the four parameters all weak decays that include in addition to tree level decays mediated by $W^{\pm}$-bosons, a vast number of one-loop induced flavour changing neutral current transitions involving gluons, photon, $W^{\pm}$, Z^0 and H^0. The latter transitions are responsible for rare decays and CP-violating decays in the SM. This important role of the CKM matrix is preserved in any extension of the SM even if more complicated extensions may contain new sources of flavour violation and CP violation.

The answer to this question is very challenging because the relevant rare and CP-violating decays have small branching ratios and are often very difficult to measure. Moreover, as hadrons are bound states of quarks and antiquarks, the determination of the CKM parameters requires in many cases a quantitative control over QCD effects at long distances where the existing non-perturbative methods are not yet satisfactory.

In spite of these difficulties, we strongly believe that this important question will be answered in this decade. This belief is based on an impressive progress in the experimental measurements in this field and on a similar progress made by theorists in perturbative and to a lesser extend non-perturbative QCD calculations. The development of various strategies for the determination of the CKM parameters, that are essentially free from hadronic uncertainties, is also an important ingredient in this progress. A recent account of these joined efforts by experimentalist and theorists is given in [3].

These lecture notes provide a rather non-technical description of the decays that are best suited for the determination of the CKM matrix. We will

A.J. Buras, CP Violation in B and K Decays: 2003, Lect. Notes Phys. **629**, 85–135 (2004)
http://www.springerlink.com/   © Springer-Verlag Berlin Heidelberg 2004

also briefly discuss the ratio ε'/ε that from the present perspective is not suited for a precise determination of the CKM matrix but is interesting on its own. There is unavoidably an overlap with our Les Houches [4], Lake Louise [5], Erice [6] and Zacatecas [7] lectures and with the reviews [8] and [9]. On the other hand new developments until the summer 2003 have been taken into account, as far as the space allowed for it, and all numerical results have been updated. Moreover the discussion of the strategies for the determination of the angles α, β and γ in the unitarity triangle goes far beyond our previous lectures.

We hope that these lecture notes will be helpful in following the new developments in this exciting field. In this respect the recent books [10–12], the working group reports [3,13–15] and most recent reviews [16] are also strongly recommended.

1.2   CKM Matrix and the Unitarity Triangle

The unitary CKM matrix [1,2] connects the *weak eigenstates* (d', s', b') and the corresponding *mass eigenstates* d, s, b:

$$\begin{pmatrix} d' \\ s' \\ b' \end{pmatrix} = \begin{pmatrix} V_{ud} & V_{us} & V_{ub} \\ V_{cd} & V_{cs} & V_{cb} \\ V_{td} & V_{ts} & V_{tb} \end{pmatrix} \begin{pmatrix} d \\ s \\ b \end{pmatrix} \equiv \hat{V}_{\mathrm{CKM}} \begin{pmatrix} d \\ s \\ b \end{pmatrix}. \tag{1}$$

Many parametrizations of the CKM matrix have been proposed in the literature. The classification of different parametrizations can be found in [17]. While the so called standard parametrization [18]

$$\hat{V}_{\mathrm{CKM}} = \begin{pmatrix} c_{12}c_{13} & s_{12}c_{13} & s_{13}e^{-i\delta} \\ -s_{12}c_{23} - c_{12}s_{23}s_{13}e^{i\delta} & c_{12}c_{23} - s_{12}s_{23}s_{13}e^{i\delta} & s_{23}c_{13} \\ s_{12}s_{23} - c_{12}c_{23}s_{13}e^{i\delta} & -s_{23}c_{12} - s_{12}c_{23}s_{13}e^{i\delta} & c_{23}c_{13} \end{pmatrix}, \tag{2}$$

with $c_{ij} = \cos\theta_{ij}$ and $s_{ij} = \sin\theta_{ij}$ $(i,j = 1,2,3)$ and the complex phase δ necessary for CP violation, should be recommended [19] for any numerical analysis, a generalization of the Wolfenstein parametrization [20] as presented in [21] is more suitable for these lectures. On the one hand it is more transparent than the standard parametrization and on the other hand it satisfies the unitarity of the CKM matrix to higher accuracy than the original parametrization in [20].

To this end we make the following change of variables in the standard parametrization (2) [21,22]

$$s_{12} = \lambda, \qquad s_{23} = A\lambda^2, \qquad s_{13}e^{-i\delta} = A\lambda^3(\varrho - i\eta) \tag{3}$$

where

$$\lambda, \qquad A, \qquad \varrho, \qquad \eta \tag{4}$$

are the Wolfenstein parameters with $\lambda \approx 0.22$ being an expansion parameter. We find then

$$V_{ud} = 1 - \frac{1}{2}\lambda^2 - \frac{1}{8}\lambda^4, \qquad V_{cs} = 1 - \frac{1}{2}\lambda^2 - \frac{1}{8}\lambda^4(1 + 4A^2), \qquad (5)$$

$$V_{tb} = 1 - \frac{1}{2}A^2\lambda^4, \qquad V_{cd} = -\lambda + \frac{1}{2}A^2\lambda^5[1 - 2(\varrho + i\eta)], \qquad (6)$$

$$V_{us} = \lambda + \mathcal{O}(\lambda^7), \qquad V_{ub} = A\lambda^3(\varrho - i\eta), \qquad V_{cb} = A\lambda^2 + \mathcal{O}(\lambda^8), \qquad (7)$$

$$V_{ts} = -A\lambda^2 + \frac{1}{2}A\lambda^4[1 - 2(\varrho + i\eta)], \qquad V_{td} = A\lambda^3(1 - \bar{\varrho} - i\bar{\eta}) \qquad (8)$$

where terms $\mathcal{O}(\lambda^6)$ and higher order terms have been neglected. A non-vanishing η is responsible for CP violation in the SM. It plays the role of δ in the standard parametrization. Finally, the barred variables in (8) are given by [21]

$$\bar{\varrho} = \varrho(1 - \frac{\lambda^2}{2}), \qquad \bar{\eta} = \eta(1 - \frac{\lambda^2}{2}). \qquad (9)$$

Now, the unitarity of the CKM-matrix implies various relations between its elements. In particular, we have

$$V_{ud}V_{ub}^* + V_{cd}V_{cb}^* + V_{td}V_{tb}^* = 0. \qquad (10)$$

The relation (10) can be represented as a "unitarity" triangle in the complex $(\bar{\varrho}, \bar{\eta})$ plane. One can construct additional five unitarity triangles [23] corresponding to other unitarity relations.

Noting that to an excellent accuracy $V_{cd}V_{cb}^*$ is real with $|V_{cd}V_{cb}^*| = A\lambda^3 + \mathcal{O}(\lambda^7)$ and rescaling all terms in (10) by $A\lambda^3$ we indeed find that the relation (10) can be represented as the triangle in the complex $(\bar{\varrho}, \bar{\eta})$ plane as shown in Fig. 1. Let us collect useful formulae related to this triangle:

- We can express $\sin(2\phi_i)$, $\phi_i = \alpha, \beta, \gamma$, in terms of $(\bar{\varrho}, \bar{\eta})$. In particular:

$$\sin(2\beta) = \frac{2\bar{\eta}(1 - \bar{\varrho})}{(1 - \bar{\varrho})^2 + \bar{\eta}^2}. \qquad (11)$$

- The lengths CA and BA are given respectively by

$$R_b \equiv \frac{|V_{ud}V_{ub}^*|}{|V_{cd}V_{cb}^*|} = \sqrt{\bar{\varrho}^2 + \bar{\eta}^2} = (1 - \frac{\lambda^2}{2})\frac{1}{\lambda}\left|\frac{V_{ub}}{V_{cb}}\right|, \qquad (12)$$

$$R_t \equiv \frac{|V_{td}V_{tb}^*|}{|V_{cd}V_{cb}^*|} = \sqrt{(1 - \bar{\varrho})^2 + \bar{\eta}^2} = \frac{1}{\lambda}\left|\frac{V_{td}}{V_{cb}}\right|. \qquad (13)$$

- The angles β and $\gamma = \delta$ of the unitarity triangle are related directly to the complex phases of the CKM-elements V_{td} and V_{ub}, respectively, through

$$V_{td} = |V_{td}|e^{-i\beta}, \qquad V_{ub} = |V_{ub}|e^{-i\gamma}. \qquad (14)$$

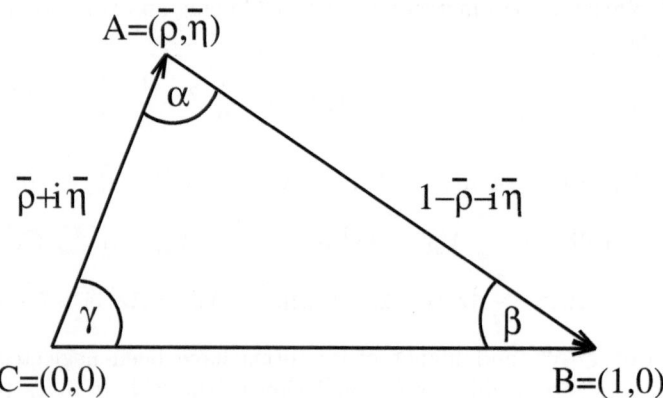

Fig. 1. Unitarity Triangle.

- The unitarity relation (10) can be rewritten as

$$R_b e^{i\gamma} + R_t e^{-i\beta} = 1 \ . \tag{15}$$

- The angle α can be obtained through the relation

$$\alpha + \beta + \gamma = 180° \ . \tag{16}$$

Formula (15) shows transparently that the knowledge of (R_t, β) allows to determine (R_b, γ) through

$$R_b = \sqrt{1 + R_t^2 - 2R_t \cos \beta}, \qquad \cot \gamma = \frac{1 - R_t \cos \beta}{R_t \sin \beta}. \tag{17}$$

Similarly, (R_t, β) can be expressed through (R_b, γ):

$$R_t = \sqrt{1 + R_b^2 - 2R_b \cos \gamma}, \qquad \cot \beta = \frac{1 - R_b \cos \gamma}{R_b \sin \gamma}. \tag{18}$$

These relations are remarkable. They imply that the knowledge of the coupling V_{td} between t and d quarks allows to deduce the strength of the corresponding coupling V_{ub} between u and b quarks and vice versa.

The triangle depicted in Fig. 1, $|V_{us}|$ and $|V_{cb}|$ give the full description of the CKM matrix. Looking at the expressions for R_b and R_t, we observe that within the SM the measurements of four CP *conserving* decays sensitive to $|V_{us}|, |V_{ub}|, |V_{cb}|$ and $|V_{td}|$ can tell us whether CP violation ($\bar{\eta} \neq 0$ or $\gamma \neq 0, \pi$) is predicted in the SM. This fact is often used to determine the angles of the unitarity triangle without the study of CP-violating quantities.

1.3   The Special Role of $|V_{us}|$, $|V_{ub}|$ and $|V_{cb}|$

What do we know about the CKM matrix and the unitarity triangle on the basis of *tree level* decays? Here the semi-leptonic K and B decays play the decisive role. The present situation can be summarized by [3]

$$|V_{us}| = \lambda = 0.2240 \pm 0.0036 \qquad |V_{cb}| = (41.5 \pm 0.8) \cdot 10^{-3}, \qquad (19)$$

$$\frac{|V_{ub}|}{|V_{cb}|} = 0.086 \pm 0.008, \qquad |V_{ub}| = (3.57 \pm 0.31) \cdot 10^{-3}. \qquad (20)$$

implying

$$A = 0.83 \pm 0.02, \qquad R_b = 0.37 \pm 0.04 . \qquad (21)$$

There is an impressive work done by theorists and experimentalists hidden behind these numbers. We refer to [3] for details. See also [19].

The information given above tells us only that the apex A of the unitarity triangle lies in the band shown in Fig. 2. While this information appears at first sight to be rather limited, it is very important for the following reason. As $|V_{us}|$, $|V_{cb}|$, $|V_{ub}|$ and consequently R_b are determined here from tree level decays, their values given above are to an excellent accuracy independent of any new physics contributions. They are universal fundamental constants valid in any extention of the SM. Therefore their precise determinations are of utmost importance.

In order to answer the question where the apex A lies on the "unitarity clock" in Fig. 2 we have to look at other decays. Most promising in this respect are the so-called "loop induced" decays and transitions and CP-violating B decays. These decays are sensitive to the angles β and γ as well as to the length R_t and measuring only one of these three quantities allows to find the unitarity triangle provided the universal R_b is known.

Of course any pair among (R_t, β, γ) is sufficient to construct the UT without any knowledge of R_b. Yet the special role of R_b among these variables lies in its universality whereas the other three variables are generally sensitive functions of possible new physics contributions. This means that assuming three generation unitarity of the CKM matrix and that the SM is a part of a bigger theory, the apex of the unitarity triangle has to be eventually placed

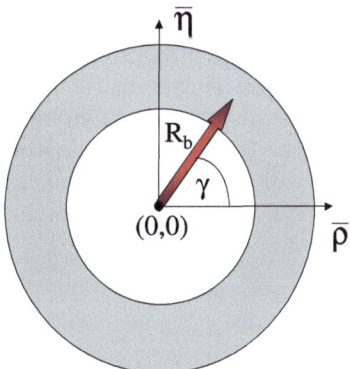

Fig. 2. "Unitarity Clock".

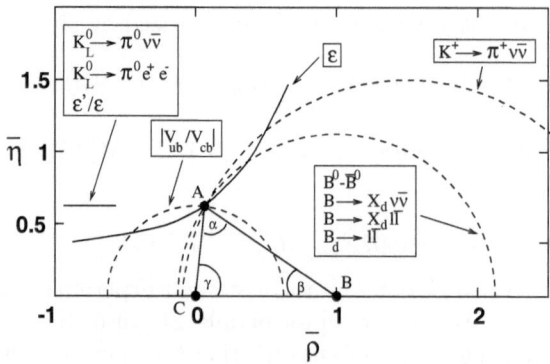

Fig. 3. The ideal Unitarity Triangle.

on the unitarity clock with the radius R_b obtained from tree level decays. That is even if using SM expressions for loop induced processes, $(\bar{\varrho}, \bar{\eta})$ would be found outside the unitarity clock, the corresponding expressions of the grander theory must include appropriate new contributions so that the apex of the unitarity triangle is shifted back to the band in Fig. 2. In the case of CP asymmetries this could be achieved by realizing that the measured angles α, β and γ are not the true angles of the unitarity triangle but sums of the true angles and new complex phases present in extentions of the SM. The better R_b is known, the thiner the band in Fig. 2 will be, selecting in this manner efficiently the correct theory. On the other hand as the the branching ratios for rare and CP-violating decays depend sensitively on the parameter A, the precise knowledge of $|V_{cb}|$ is also very important.

1.4 Grand Picture

The apex $(\bar{\varrho}, \bar{\eta})$ of the UT can be efficiently hunted by means of rare and CP violating transitions as shown in Fig. 3. Moreover the angles of this triangle can be measured in CP asymmetries in B-decays and using other strategies. This picture could describe in principle the reality in the year 2012, my retirement year, if the SM is the whole story. On the other hand in the presence of significant new physics contributions, the use of the SM expressions for rare and CP violating transitions in question, combined with future precise measurements, may result in curves which do not cross each other at a single point in the $(\bar{\varrho}, \bar{\eta})$ plane. This would be truly exciting and most of us hope that this will turn out to be the case. In order to be able to draw such thin curves as in Fig. 3, not only experiments but also the theory has to be under control.

1.5 Theoretical Framework

The present framework for weak decays is based on the operator product expansion (OPE) that allows to separate short (SD) and long (LD) distance contributions to weak amplitudes and on the renormalization group (RG) methods that allow to sum large logarithms $\log \mu_{SD}/\mu_{LD}$ to all orders in perturbation theory. The full exposition of these methods can be found in [4,8].

The OPE allows to write the effective weak Hamiltonian simply as follows

$$\mathcal{H}_{eff} = \frac{G_F}{\sqrt{2}} \sum_i V_{\text{CKM}}^i C_i(\mu) Q_i . \tag{22}$$

Here G_F is the Fermi constant and Q_i are the relevant local operators which govern the decays in question. They are built out of quark and lepton fields. The Cabibbo-Kobayashi-Maskawa factors V_{CKM}^i [1,2] and the Wilson coefficients $C_i(\mu)$ describe the strength with which a given operator enters the Hamiltonian. The latter coefficients can be considered as scale dependent "couplings" related to "vertices" Q_i and as discussed below can be calculated using perturbative methods as long as μ is not too small.

An amplitude for a decay of a given meson $M = K, B, ..$ into a final state $F = \pi\nu\bar{\nu}$, $\pi\pi$, DK is then simply given by

$$A(M \to F) = \langle F|\mathcal{H}_{eff}|M\rangle = \frac{G_F}{\sqrt{2}} \sum_i V_{CKM}^i C_i(\mu) \langle F|Q_i(\mu)|M\rangle, \tag{23}$$

where $\langle F|Q_i(\mu)|M\rangle$ are the matrix elements of Q_i between M and F, evaluated at the renormalization scale μ.

The essential virtue of OPE is this one. It allows to separate the problem of calculating the amplitude $A(M \to F)$ into two distinct parts: the *short distance* (perturbative) calculation of the coefficients $C_i(\mu)$ and the *long-distance* (generally non-perturbative) calculation of the matrix elements $\langle Q_i(\mu)\rangle$. The scale μ separates, roughly speaking, the physics contributions into short distance contributions contained in $C_i(\mu)$ and the long distance contributions contained in $\langle Q_i(\mu)\rangle$. Thus C_i include the top quark contributions and contributions from other heavy particles such as W-, Z-bosons and charged Higgs particles or supersymmetric particles in the supersymmetric extensions of the SM. Consequently $C_i(\mu)$ depend generally on m_t and also on the masses of new particles if extensions of the SM are considered. This dependence can be found by evaluating so-called *box* and *penguin* diagrams with full W-, Z-, top- and new particles exchanges and *properly* including short distance QCD effects. The latter govern the μ-dependence of $C_i(\mu)$.

The value of μ can be chosen arbitrarily but the final result must be μ-independent. Therefore the μ-dependence of $C_i(\mu)$ has to cancel the μ-dependence of $\langle Q_i(\mu)\rangle$. The same comments apply to the renormalization scheme dependence of $C_i(\mu)$ and $\langle Q_i(\mu)\rangle$.

Now due to the fact that for low energy processes the appropriate scale $\mu \ll M_{W,Z}$, m_t, large logarithms $\ln M_W/\mu$ compensate in the evaluation of $C_i(\mu)$ the smallness of the QCD coupling constant α_s and terms $\alpha_s^n(\ln M_W/\mu)^n$, $\alpha_s^n(\ln M_W/\mu)^{n-1}$ etc. have to be resummed to all orders in

α_s before a reliable result for C_i can be obtained. This can be done very efficiently by means of the renormalization group methods. The resulting *renormalization group improved* perturbative expansion for $C_i(\mu)$ in terms of the effective coupling constant $\alpha_s(\mu)$ does not involve large logarithms and is more reliable. The related technical issues are discussed in detail in [4] and [8]. It should be emphasized that by 2003 the next-to-leading (NLO) QCD and QED corrections to all relevant weak decay processes in the SM are known.

Clearly, in order to calculate the amplitude $A(M \to F)$ the matrix elements $\langle Q_i(\mu) \rangle$ have to be evaluated. Since they involve long distance contributions one is forced in this case to use non-perturbative methods such as lattice calculations, the $1/N$ expansion (N is the number of colours), QCD sum rules, hadronic sum rules and chiral perturbation theory. In the case of B-meson decays, the *Heavy Quark Effective Theory* (HQET) and *Heavy Quark Expansions* (HQE) also turn out to be useful tools. However, all these non-perturbative methods have some limitations. Consequently the dominant theoretical uncertainties in the decay amplitudes reside in the matrix elements $\langle Q_i(\mu) \rangle$ and non-perturbative parameters present in HQET and HQE. These issues are reviewed in [3].

The fact that in many cases the matrix elements $\langle Q_i(\mu) \rangle$ cannot be reliably calculated at present, is very unfortunate. The main goals of the experimental studies of weak decays is the determination of the CKM factors V_{CKM} and the search for the physics beyond the SM. Without a reliable estimate of $\langle Q_i(\mu) \rangle$ these goals cannot be achieved unless these matrix elements can be determined experimentally or removed from the final measurable quantities by taking suitable ratios and combinations of decay amplitudes or branching ratios. We will encounter many examples in these lectures. Flavour symmetries like $SU(2)_F$ and $SU(3)_F$ relating various matrix elements can be useful in this respect, provided flavour symmetry breaking effects can be reliably calculated. A recent progress in the calculation of $\langle Q_i(\mu) \rangle$ relevant for non-leptonic B decays can be very helpful here as discussed in Sect. 5.

After these general remarks let us be more specific about the structure of (23) by considering the simplest class of models in which all flavour violating and CP-violating transition are governed by the CKM matrix and the only relevant local operators are the ones that are relevant in the SM. We will call this scenario "Minimal Flavour Violation" (MFV) [24] being aware of the fact that for some authors MFV means a more general framework in which also new operators can give significant contributions. See for instance the recent discussions in [25,26]. In the MFV models, as defined in [24], the formula (23) can be written as follows

$$A(\text{Decay}) = \sum_i B_i \eta^i_{\text{QCD}} V^i_{\text{CKM}} F^i, \qquad F^i = F^i_{\text{SM}} + F^i_{\text{New}} \qquad (24)$$

with F^i_{SM} and F^i_{New} being real.

Here the non-perturbative parameters B_i represent the matrix elements of local operators present in the SM. For instance in the case of $K^0 - \bar{K}^0$ mixing, the matrix element of the operator $\bar{s}\gamma_\mu(1-\gamma_5)d \otimes \bar{s}\gamma^\mu(1-\gamma_5)d$ is represented by the parameter $\hat{B}_K$. There are other non-perturbative parameters in the SM that represent matrix elements of operators Q_i with different colour and Dirac structures. The objects η_{QCD}^i are the QCD factors resulting from RG-analysis of the corresponding operators and F_{SM}^i stand for the so-called Inami-Lim functions [27] that result from the calculations of various box and penguin diagrams. They depend on the top-quark mass. V_{CKM}^i are the CKM-factors we want to determine.

The important point is that in all MFV models B_i and η_{QCD}^i are the same as in the SM and the only place where the new physics enters are the new short distance functions F_{New}^i that depend on the new parameters in the extensions of the SM like the masses of charginos, squarks, charged Higgs particles and $\tan\beta = v_2/v_1$ in the MSSM. These new particles enter the new box and penguin diagrams. Strictly speaking at the NLO level the QCD corrections to the new diagrams at scales larger than $\mathcal{O}(M_\mathrm{W})$ may differ from the corresponding corrections in the SM but this, generally small, difference can be absorbed into F_{New}^i so that η_{QCD}^i are the QCD corrections calculated in the SM. Indeed, the QCD corrections at scales lower than $\mathcal{O}(M_\mathrm{W})$ are related to the renormalization of the local operators that are common to all models in this class.

In more complicated extensions of the SM new operators (Dirac structures) that are either absent or very strongly suppressed in the SM, can become important. Moreover new sources of flavour and CP violation beyond the CKM matrix, including new complex phases, could be present. A general master formula describing such contributions is given in [28].

Finally, let me give some arguments why our definition of MFV models is phenomenologically useful. With a simple formula like (24) it is possible to derive a number of relations that are independent of the parameters specific to a given MFV models. Consequently, any violation of these relations will signal the presence of new local operators and/or new complex phases that are necessary to describe the data. We will return to this point in Sect. 7.

2 Particle-Antiparticle Mixing and Various Types of CP Violation

2.1 Preliminaries

Let us next discuss the formalism of particle–antiparticle mixing and CP violation. Much more elaborate discussion can be found in two books [11,12]. We will concentrate here on $K^0 - \bar{K}^0$ mixing, $B_{d,s}^0 - \bar{B}_{d,s}^0$ mixings and CP violation in K-meson and B-meson decays. Due to GIM mechanism [29] the phenomena discussed in this section appear first at the one–loop level and as such they are sensitive measures of the top quark couplings $V_{ti}(i = d, s, b)$ and in particular of the phase $\delta = \gamma$. They allow then to construct the unitarity triangle as explicitly demonstrated in Sect. 4.

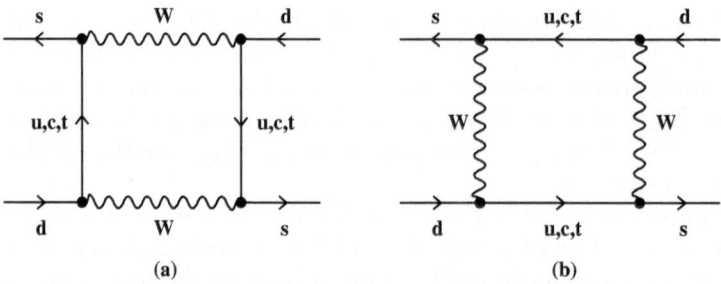

Fig. 4. Box diagrams contributing to $K^0 - \bar{K}^0$ mixing in the SM.

2.2   Express Review of $K^0 - \bar{K}^0$ Mixing

$K^0 = (\bar{s}d)$ and $\bar{K}^0 = (s\bar{d})$ are flavour eigenstates which in the SM may mix via weak interactions through the box diagrams in Fig. 4. We will choose the phase conventions so that

$$CP|K^0\rangle = -|\bar{K}^0\rangle, \qquad CP|\bar{K}^0\rangle = -|K^0\rangle. \tag{25}$$

In the absence of mixing the time evolution of $|K^0(t)\rangle$ is given by

$$|K^0(t)\rangle = |K^0(0)\rangle \exp(-iHt), \qquad H = M - i\frac{\Gamma}{2}, \tag{26}$$

where M is the mass and Γ the width of K^0. Similar formula exists for $\bar{K}^0$.

On the other hand, in the presence of flavour mixing the time evolution of the $K^0 - \bar{K}^0$ system is described by

$$i\frac{d\psi(t)}{dt} = \hat{H}\psi(t) \qquad \psi(t) = \begin{pmatrix} |K^0(t)\rangle \\ |\bar{K}^0(t)\rangle \end{pmatrix} \tag{27}$$

where

$$\hat{H} = \hat{M} - i\frac{\hat{\Gamma}}{2} = \begin{pmatrix} M_{11} - i\frac{\Gamma_{11}}{2} & M_{12} - i\frac{\Gamma_{12}}{2} \\ M_{21} - i\frac{\Gamma_{21}}{2} & M_{22} - i\frac{\Gamma_{22}}{2} \end{pmatrix} \tag{28}$$

with $\hat{M}$ and $\hat{\Gamma}$ being hermitian matrices having positive (real) eigenvalues in analogy with M and Γ. M_{ij} and Γ_{ij} are the transition matrix elements from virtual and physical intermediate states respectively. Using

$$M_{21} = M_{12}^*, \qquad \Gamma_{21} = \Gamma_{12}^*, \qquad \text{(hermiticity)} \tag{29}$$

$$M_{11} = M_{22} \equiv M, \qquad \Gamma_{11} = \Gamma_{22} \equiv \Gamma, \qquad \text{(CPT)} \tag{30}$$

we have

$$\hat{H} = \begin{pmatrix} M - i\frac{\Gamma}{2} & M_{12} - i\frac{\Gamma_{12}}{2} \\ M_{12}^* - i\frac{\Gamma_{12}^*}{2} & M - i\frac{\Gamma}{2} \end{pmatrix} . \tag{31}$$

Diagonalizing (27) we find:

Eigenstates:

$$K_{L,S} = \frac{(1+\bar{\varepsilon})K^0 \pm (1-\bar{\varepsilon})\bar{K}^0}{\sqrt{2(1+ | \bar{\varepsilon} |^2)}} \tag{32}$$

where $\bar{\varepsilon}$ is a small complex parameter given by

$$\frac{1-\bar{\varepsilon}}{1+\bar{\varepsilon}} = \sqrt{\frac{M_{12}^* - i\frac{1}{2}\Gamma_{12}^*}{M_{12} - i\frac{1}{2}\Gamma_{12}}} = \frac{2M_{12}^* - i\Gamma_{12}^*}{\Delta M - i\frac{1}{2}\Delta\Gamma} \equiv r\exp(i\kappa) . \tag{33}$$

with $\Delta\Gamma$ and ΔM given below.

Eigenvalues:

$$M_{L,S} = M \pm \mathrm{Re}Q \qquad \Gamma_{L,S} = \Gamma \mp 2\mathrm{Im}Q \tag{34}$$

where

$$Q = \sqrt{(M_{12} - i\frac{1}{2}\Gamma_{12})(M_{12}^* - i\frac{1}{2}\Gamma_{12}^*)}. \tag{35}$$

Consequently we have

$$\Delta M = M_L - M_S = 2\mathrm{Re}Q , \qquad \Delta\Gamma = \Gamma_L - \Gamma_S = -4\mathrm{Im}Q. \tag{36}$$

It should be noted that the mass eigenstates K_S and K_L differ from the CP eigenstates

$$K_1 = \frac{1}{\sqrt{2}}(K^0 - \bar{K}^0), \qquad CP|K_1\rangle = |K_1\rangle , \tag{37}$$

$$K_2 = \frac{1}{\sqrt{2}}(K^0 + \bar{K}^0), \qquad CP|K_2\rangle = -|K_2\rangle , \tag{38}$$

by a small admixture of the other CP eigenstate:

$$K_S = \frac{K_1 + \bar{\varepsilon}K_2}{\sqrt{1+ | \bar{\varepsilon} |^2}}, \qquad K_L = \frac{K_2 + \bar{\varepsilon}K_1}{\sqrt{1+ | \bar{\varepsilon} |^2}} . \tag{39}$$

Since $\bar{\varepsilon}$ is $\mathcal{O}(10^{-3})$, one has to a very good approximation:

$$\Delta M_K = 2\mathrm{Re}M_{12}, \qquad \Delta\Gamma_K = 2\mathrm{Re}\Gamma_{12} , \tag{40}$$

where we have introduced the subscript K to stress that these formulae apply only to the $K^0 - \bar{K}^0$ system.

The $K_L - K_S$ mass difference is experimentally measured to be [19]

$$\Delta M_K = M(K_L) - M(K_S) = (3.490 \pm 0.006) \cdot 10^{-15}\,\mathrm{GeV}\,. \qquad (41)$$

In the SM roughly 80% of the measured ΔM_K is described by the real parts of the box diagrams with charm quark and top quark exchanges, whereby the contribution of the charm exchanges is by far dominant. The remaining 20% of the measured ΔM_K is attributed to long distance contributions which are difficult to estimate [30]. Further information with the relevant references can be found in [31]. The situation with $\Delta \Gamma_K$ is rather different. It is fully dominated by long distance effects. Experimentally one has $\Delta \Gamma_K \approx -2\Delta M_K$.

Generally to observe CP violation one needs an interference between various amplitudes that carry complex phases. As these phases are obviously convention dependent, the CP-violating effects depend only on the differences of these phases. In particular the parameter $\bar{\varepsilon}$ depends on the phase convention chosen for K^0 and $\bar{K}^0$. Therefore it may not be taken as a physical measure of CP violation. On the other hand Re $\bar{\varepsilon}$ and r, defined in (33) are independent of phase conventions. In fact the departure of r from 1 measures CP violation in the $K^0 - \bar{K}^0$ mixing:

$$r = 1 + \frac{2|\Gamma_{12}|^2}{4|M_{12}|^2 + |\Gamma_{12}|^2}\mathrm{Im}\left(\frac{M_{12}}{\Gamma_{12}}\right) \approx 1 - \mathrm{Im}\left(\frac{\Gamma_{12}}{M_{12}}\right)\,. \qquad (42)$$

This type of CP violation can be best isolated in semi-leptonic decays of the K_L meson. The non-vanishing asymmetry $a_{\mathrm{SL}}(K_L)$:

$$\frac{\Gamma(K_L \to \pi^- e^+ \nu_e) - \Gamma(K_L \to \pi^+ e^- \bar{\nu}_e)}{\Gamma(K_L \to \pi^- e^+ \nu_e) + \Gamma(K_L \to \pi^+ e^- \bar{\nu}_e)} = \left(\mathrm{Im}\frac{\Gamma_{12}}{M_{12}}\right)_K = 2\mathrm{Re}\bar{\varepsilon} \qquad (43)$$

signals this type of CP violation. Note that $a_{\mathrm{SL}}(\mathrm{K_L})$ is determined purely by the quantities related to $K^0 - \bar{K}^0$ mixing. Specifically, it measures the difference between the phases of Γ_{12} and M_{12}.

That a non–vanishing $a_{\mathrm{SL}}(K_L)$ is indeed a signal of CP violation can also be understood in the following manner. K_L, that should be a CP eigenstate K_2 in the case of CP conservation, decays into CP conjugate final states with different rates. As Re$\bar{\varepsilon} > 0$, K_L prefers slightly to decay into $\pi^- e^+ \nu_e$ than $\pi^+ e^- \bar{\nu}_e$. This would not be possible in a CP-conserving world.

2.3   The First Look at ε and ε'

Since a two pion final state is CP even while a three pion final state is CP odd, K_S and K_L preferably decay to 2π and 3π, respectively via the following CP-conserving decay modes:

$$K_L \to 3\pi \ \ (\text{via K}_2), \qquad K_S \to 2\pi \ \ (\text{via K}_1). \qquad (44)$$

This difference is responsible for the large disparity in their life-times. A factor of 579. However, K_L and K_S are not CP eigenstates and may decay with small branching fractions as follows:

$$K_L \to 2\pi \quad (\text{via } K_1), \qquad K_S \to 3\pi \quad (\text{via } K_2). \tag{45}$$

This violation of CP is called *indirect* as it proceeds not via explicit breaking of the CP symmetry in the decay itself but via the admixture of the CP state with opposite CP parity to the dominant one. The measure for this indirect CP violation is defined as (I=isospin)

$$\varepsilon \equiv \frac{A(K_L \to (\pi\pi)_{I=0})}{A(K_S \to (\pi\pi)_{I=0})}. \tag{46}$$

Following the derivation in [32] one finds

$$\varepsilon = \bar{\varepsilon} + i\xi = \frac{\exp(i\pi/4)}{\sqrt{2}\Delta M_K} \left(\mathrm{Im}M_{12} + 2\xi\mathrm{Re}M_{12}\right), \qquad \xi = \frac{\mathrm{Im}A_0}{\mathrm{Re}A_0}. \tag{47}$$

The phase convention dependence of ξ cancels the one of $\bar{\varepsilon}$ so that ε is free from this dependence. The isospin amplitude A_0 is defined below.

The important point in the definition (46) is that only the transition to $(\pi\pi)_{I=0}$ enters. The transition to $(\pi\pi)_{I=2}$ is absent. This allows to remove a certain type of CP violation that originates in decays only. Yet as $\varepsilon \neq \bar{\varepsilon}$ and only $\mathrm{Re}\varepsilon = \mathrm{Re}\bar{\varepsilon}$, it is clear that ε includes a type of CP violation represented by $\mathrm{Im}\varepsilon$ which is absent in the semileptonic asymmetry (43). We will identify this type of CP violation in Sect. 2.7, where a more systematic classification of different types of CP violation will be given.

While *indirect* CP violation reflects the fact that the mass eigenstates are not CP eigenstates, so-called *direct* CP violation is realized via a direct transition of a CP odd to a CP even state: $K_2 \to \pi\pi$. A measure of such a direct CP violation in $K_L \to \pi\pi$ is characterized by a complex parameter ε' defined as

$$\varepsilon' \equiv \frac{1}{\sqrt{2}} \left(\frac{A_{2,L}}{A_{0,S}} - \frac{A_{2,S}}{A_{0,S}} \frac{A_{0,L}}{A_{0,S}} \right) \tag{48}$$

where $A_{I,L} \equiv A(K_L \to (\pi\pi)_I)$ and $A_{I,S} \equiv A(K_S \to (\pi\pi)_I)$.

This time the transitions to $(\pi\pi)_{I=0}$ and $(\pi\pi)_{I=2}$ are included which allows to study CP violation in the decay itself. We will discuss this issue in general terms in Sect. 2.7. It is useful to cast (48) into

$$\varepsilon' = \frac{1}{\sqrt{2}}\mathrm{Im}\left(\frac{A_2}{A_0}\right)\exp(i\Phi_{\varepsilon'}), \qquad \Phi_{\varepsilon'} = \frac{\pi}{2} + \delta_2 - \delta_0, \tag{49}$$

where the isospin amplitudes A_I in $K \to \pi\pi$ decays are introduced through

$$A(K^+ \to \pi^+\pi^0) = \sqrt{\frac{3}{2}}A_2 e^{i\delta_2}, \tag{50}$$

$$A(K^0 \to \pi^+\pi^-) = \sqrt{\frac{2}{3}} A_0 e^{i\delta_0} + \sqrt{\frac{1}{3}} A_2 e^{i\delta_2} , \tag{51}$$

$$A(K^0 \to \pi^0\pi^0) = \sqrt{\frac{2}{3}} A_0 e^{i\delta_0} - 2\sqrt{\frac{1}{3}} A_2 e^{i\delta_2} . \tag{52}$$

Here the subscript $I = 0, 2$ denotes states with isospin 0, 2 equivalent to $\Delta I = 1/2$ and $\Delta I = 3/2$ transitions, respectively, and $\delta_{0,2}$ are the corresponding strong phases. The weak CKM phases are contained in A_0 and A_2. The isospin amplitudes A_I are complex quantities which depend on phase conventions. On the other hand, ε' measures the difference between the phases of A_2 and A_0 and is a physical quantity. The strong phases $\delta_{0,2}$ can be extracted from $\pi\pi$ scattering. Then $\Phi_{\varepsilon'} \approx \pi/4$. See [33] for more details.

Experimentally ε and ε' can be found by measuring the ratios

$$\eta_{00} = \frac{A(K_{\rm L} \to \pi^0\pi^0)}{A(K_{\rm S} \to \pi^0\pi^0)}, \qquad \eta_{+-} = \frac{A(K_{\rm L} \to \pi^+\pi^-)}{A(K_{\rm S} \to \pi^+\pi^-)}. \tag{53}$$

Indeed, assuming ε and ε' to be small numbers one finds

$$\eta_{00} = \varepsilon - \frac{2\varepsilon'}{1 - \sqrt{2}\omega}, \qquad \eta_{+-} = \varepsilon + \frac{\varepsilon'}{1 + \omega/\sqrt{2}} \tag{54}$$

where $\omega = {\rm Re}A_2/{\rm Re}A_0 = 0.045$. In the absence of direct CP violation $\eta_{00} = \eta_{+-}$. The ratio ε'/ε can then be measured through

$$\mathrm{Re}(\varepsilon'/\varepsilon) = \frac{1}{6(1 + \omega/\sqrt{2})} \left(1 - \left| \frac{\eta_{00}}{\eta_{+-}} \right|^2 \right) . \tag{55}$$

2.4   Basic Formula for ε

With all this information at hand one can derive a formula for ε which can be efficiently used in pheneomenological applications. As this derivation has been presented in detail in [6], we will be very brief here.

Calculating the box diagrams of Fig. 4 and including leading and next-to-leading QCD corrections one finds

$$M_{12} = D_\varepsilon \left[\lambda_c^{*2}\eta_1 S_0(x_c) + \lambda_t^{*2}\eta_2 S_0(x_t) + 2\lambda_c^*\lambda_t^*\eta_3 S_0(x_c, x_t) \right], \tag{56}$$

$$D_\varepsilon = \frac{G_{\rm F}^2}{12\pi^2} F_K^2 \hat{B}_K m_K M_{\rm W}^2 \tag{57}$$

where $F_K = 160$ MeV is the K-meson decay constant and m_K the K-meson mass. Next, the renormalization group invariant parameter $\hat{B}_K$ is defined by

$$\hat{B}_K = B_K(\mu) \left[\alpha_s^{(3)}(\mu) \right]^{-2/9} \left[1 + \frac{\alpha_s^{(3)}(\mu)}{4\pi} J_3 \right], \tag{58}$$

$$\langle \bar{K}^0|(\bar{s}d)_{V-A}(\bar{s}d)_{V-A}|K^0\rangle \equiv \frac{8}{3}B_K(\mu)F_K^2 m_K^2 \tag{59}$$

where $\alpha_s^{(3)}$ is the strong coupling constant in an effective three flavour theory and $J_3 = 1.895$ in the NDR scheme [34]. The CKM factors are given by $\lambda_i = V_{is}^* V_{id}$ and the functions S_0 by $(x_i = m_i^2/M_W^2)$

$$S_0(x_t) = 2.39 \left(\frac{m_t}{167\,\text{GeV}}\right)^{1.52}, \qquad S_0(x_c) = x_c, \tag{60}$$

$$S_0(x_c, x_t) = x_c \left[\ln \frac{x_t}{x_c} - \frac{3x_t}{4(1-x_t)} - \frac{3x_t^2 \ln x_t}{4(1-x_t)^2}\right]. \tag{61}$$

Short-distance NLO QCD effects are described through the correction factors η_1, η_2, η_3 [31,34–36]:

$$\eta_1 = (1.32 \pm 0.32) \left[\frac{1.30\,\text{GeV}}{m_c(m_c)}\right]^{1.1}, \ \eta_2 = 0.57 \pm 0.01, \ \eta_3 = 0.47 \pm 0.05 . \tag{62}$$

To proceed further we neglect the last term in (47) as it constitutes at most a 2 % correction to ε. This is justified in view of other uncertainties, in particular those connected with $\hat{B}_K$. Inserting (56) into (47) we find

$$\varepsilon = C_\varepsilon \hat{B}_K \text{Im}\lambda_t \left\{\text{Re}\lambda_c \left[\eta_1 S_0(x_c) - \eta_3 S_0(x_c, x_t)\right] - \text{Re}\lambda_t \eta_2 S_0(x_t)\right\} e^{i\pi/4} , \tag{63}$$

where the numerical constant C_ε is given by

$$C_\varepsilon = \frac{G_F^2 F_K^2 m_K M_W^2}{6\sqrt{2}\pi^2 \Delta M_K} = 3.837 \cdot 10^4 . \tag{64}$$

Comparing (63) with the experimental value for ε [19]

$$\varepsilon_{exp} = (2.280 \pm 0.013) \cdot 10^{-3} \, \exp i\Phi_\varepsilon, \qquad \Phi_\varepsilon = \frac{\pi}{4}, \tag{65}$$

one obtains a constraint on the unitarity triangle in Fig. 1. See Sect. 3.

2.5 Express Review of $B_{d,s}^0$-$\bar{B}_{d,s}^0$ Mixing

The flavour eigenstates in this case are

$$B_d^0 = (\bar{b}d), \qquad \bar{B}_d^0 = (b\bar{d}), \qquad B_s^0 = (\bar{b}s), \qquad \bar{B}_s^0 = (b\bar{s}) . \tag{66}$$

They mix via the box diagrams in Fig. 4 with s replaced by b in the case of B_d^0-$\bar{B}_d^0$ mixing. In the case of B_s^0-$\bar{B}_s^0$ mixing also d has to be replaced by s.

Dropping the subscripts (d, s) for a moment, it is customary to denote the mass eigenstates by

$$B_H = pB^0 + q\bar{B}^0, \qquad B_L = pB^0 - q\bar{B}^0, \tag{67}$$

$$p = \frac{1 + \bar{\varepsilon}_B}{\sqrt{2(1 + |\bar{\varepsilon}_B|^2)}}, \qquad q = \frac{1 - \bar{\varepsilon}_B}{\sqrt{2(1 + |\bar{\varepsilon}_B|^2)}}, \qquad (68)$$

with $\bar{\varepsilon}_B$ corresponding to $\bar{\varepsilon}$ in the $K^0 - \bar{K}^0$ system. Here "H" and "L" denote *Heavy* and *Light* respectively. As in the $B^0 - \bar{B}^0$ system one has $\Delta\Gamma \ll \Delta M$, it is more suitable to distinguish the mass eigenstates by their masses than the corresponding life-times.

The strength of the $B^0_{d,s} - \bar{B}^0_{d,s}$ mixings is described by the mass differences

$$\Delta M_{d,s} = M_H^{d,s} - M_L^{d,s} . \qquad (69)$$

In contrast to ΔM_K , in this case the long distance contributions are estimated to be very small and $\Delta M_{d,s}$ is very well approximated by the relevant box diagrams. Moreover, due $m_{u,c} \ll m_t$ only the top sector is relevant.

$\Delta M_{d,s}$ can be expressed in terms of the off-diagonal element in the neutral B-meson mass matrix by using the formulae developed previously for the K-meson system. One finds

$$\Delta M_q = 2|M_{12}^{(q)}|, \qquad \Delta\Gamma_q = 2\frac{\mathrm{Re}(M_{12}\Gamma_{12}^*)}{|M_{12}|} \ll \Delta M_q, \qquad q = d, s. \qquad (70)$$

These formulae differ from (40) because in the B-system $\Gamma_{12} \ll M_{12}$.

We also have

$$\frac{q}{p} = \frac{2M_{12}^* - i\Gamma_{12}^*}{\Delta M - i\frac{1}{2}\Delta\Gamma} = \frac{M_{12}^*}{|M_{12}|}\left[1 - \frac{1}{2}\mathrm{Im}\left(\frac{\Gamma_{12}}{M_{12}}\right)\right] \qquad (71)$$

where higher order terms in the small quantity Γ_{12}/M_{12} have been neglected. As $\mathrm{Im}(\Gamma_{12}/M_{12}) < \mathcal{O}(10^{-3})$,

- The semileptonic asymmetry $a_{\mathrm{SL}}(B)$ discussed a few pages below is even smaller than $a_{\mathrm{SL}}(K_L)$. Typically $\mathcal{O}(10^{-4})$. These are bad news.
- The ratio q/p is a pure phase to an excellent approximation. These are very good news as we will see below.

Inspecting the relevant box diagrams we find

$$(M_{12}^*)_d \propto (V_{td}V_{tb}^*)^2 , \qquad (M_{12}^*)_s \propto (V_{ts}V_{tb}^*)^2 . \qquad (72)$$

Now, from Sect. 1 we know that

$$V_{td} = |V_{td}|e^{-i\beta}, \qquad V_{ts} = -|V_{ts}|e^{-i\beta_s} \qquad (73)$$

with $\beta_s = \mathcal{O}(10^{-2})$. Consequently to an excellent approximation

$$\left(\frac{q}{p}\right)_{d,s} = e^{i2\phi_M^{d,s}}, \qquad \phi_M^d = -\beta, \qquad \phi_M^s = -\beta_s, \qquad (74)$$

with $\phi_M^{d,s}$ given entirely by the weak phases in the CKM matrix.

2.6   Basic Formulae for $\Delta M_{d,s}$

The formulae for $\Delta M_{d,s}$ have been derived in [6] with the result

$$\Delta M_q = \frac{G_F^2}{6\pi^2}\eta_B m_{B_q}(\hat{B}_{B_q}F_{B_q}^2)M_W^2 S_0(x_t)|V_{tq}|^2, \tag{75}$$

where F_{B_q} is the B_q-meson decay constant, $\hat{B}_q$ renormalization group invariant parameters defined in analogy to (58) and (59) and η_B stands for short distance QCD corrections [34,37]

$$\eta_B = 0.55 \pm 0.01. \tag{76}$$

Using (75) we obtain two useful formulae

$$\Delta M_d = 0.50/\text{ps} \cdot \left[\frac{\sqrt{\hat{B}_{B_d}}F_{B_d}}{230\,\text{MeV}}\right]^2 \left[\frac{\overline{m}_t(m_t)}{167\,\text{GeV}}\right]^{1.52} \left[\frac{|V_{td}|}{7.8\cdot 10^{-3}}\right]^2 \left[\frac{\eta_B}{0.55}\right] \tag{77}$$

and

$$\Delta M_s = 17.2/\text{ps} \cdot \left[\frac{\sqrt{\hat{B}_{B_s}}F_{B_s}}{260\,\text{MeV}}\right]^2 \left[\frac{\overline{m}_t(m_t)}{167\,\text{GeV}}\right]^{1.52} \left[\frac{|V_{ts}|}{0.040}\right]^2 \left[\frac{\eta_B}{0.55}\right]. \tag{78}$$

2.7   Classification of CP Violation

Preliminaries. We have mentioned in Sect. 1 that due to the presence of hadronic matrix elements, various decay amplitudes contain large theoretical uncertainties. It is of interest to investigate which measurements of CP-violating effects do not suffer from hadronic uncertainties. To this end it is useful to make a classification of CP-violating effects that is more transparent than the division into the *indirect* and *direct* CP violation considered so far. A nice detailed presentation has been given by Nir [16].

Generally complex phases may enter particle–antiparticle mixing and the decay process itself. It is then natural to consider three types of CP violation:

- CP Violation in Mixing
- CP Violation in Decay
- CP Violation in the Interference of Mixing and Decay

As the phases in mixing and decay are convention dependent, the CP-violating effects depend only on the differences of these phases. This is clearly seen in the classification given below.

CP Violation in Mixing. This type of CP violation can be best isolated in semi-leptonic decays of neutral B and K mesons. We have discussed the asymmetry $a_{SL}(K_L)$ before. In the case of B decays the non-vanishing asymmetry $a_{SL}(B)$ (we suppress the indices (d, s)),

$$\frac{\Gamma(\bar{B}^0(t) \to l^+\nu X) - \Gamma(B^0(t) \to l^-\bar{\nu}X)}{\Gamma(\bar{B}^0(t) \to l^+\nu X) + \Gamma(B^0(t) \to l^-\bar{\nu}X)} = \frac{1 - |q/p|^4}{1 + |q/p|^4} = \left(\mathrm{Im}\frac{\Gamma_{12}}{M_{12}}\right)_B \quad (79)$$

signals this type of CP violation. Here $\bar{B}^0(0) = \bar{B}^0$, $B^0(0) = B^0$. For $t \neq 0$ the formulae analogous to (27) should be used. Note that the final states in (79) contain "wrong charge" leptons and can only be reached in the presence of $B^0 - \bar{B}^0$ mixing. That is one studies effectively the difference between the rates for $\bar{B}^0 \to B^0 \to l^+\nu X$ and $B^0 \to \bar{B}^0 \to l^-\bar{\nu}X$. As the phases in the transitions $B^0 \to \bar{B}^0$ and $\bar{B}^0 \to B^0$ differ from each other, a non-vanishing CP asymmetry follows. Specifically $a_{\mathrm{SL}}(B)$ measures the difference between the phases of Γ_{12} and M_{12}.

As M_{12} and in particular Γ_{12} suffer from large hadronic uncertainties, no precise extraction of CP-violating phases from this type of CP violation can be expected. Moreover as q/p is almost a pure phase, see (71) and (74), the asymmetry is very small and very difficult to measure.

CP Violation in Decay. This type of CP violation is best isolated in charged B and charged K decays as mixing effects do not enter here. However, it can also be measured in neutral B and K decays. The relevant asymmetry is given by

$$a_{f\pm}^{\mathrm{decay}} = \frac{\Gamma(B^+ \to f^+) - \Gamma(B^- \to f^-)}{\Gamma(B^+ \to f^+) + \Gamma(B^- \to f^-)} = \frac{1 - |\bar{A}_{f-}/A_{f+}|^2}{1 + |\bar{A}_{f-}/A_{f+}|^2} \quad (80)$$

where

$$A_{f+} = \langle f^+|\mathcal{H}^{\mathrm{weak}}|B^+\rangle, \qquad \bar{A}_{f-} = \langle f^-|\mathcal{H}^{\mathrm{weak}}|B^-\rangle . \quad (81)$$

For this asymmetry to be non-zero one needs at least two different contributions with different *weak* (ϕ_i) and *strong* (δ_i) phases. These could be for instance two tree diagrams, two penguin diagrams or one tree and one penguin. Indeed writing the decay amplitude A_{f+} and its CP conjugate $\bar{A}_{f-}$ as

$$A_{f+} = \sum_{i=1,2} A_i e^{i(\delta_i + \phi_i)}, \qquad \bar{A}_{f-} = \sum_{i=1,2} A_i e^{i(\delta_i - \phi_i)}, \quad (82)$$

with A_i being real, one finds

$$a_{f\pm}^{\mathrm{decay}} = \frac{-2A_1 A_2 \sin(\delta_1 - \delta_2)\sin(\phi_1 - \phi_2)}{A_1^2 + A_2^2 + 2A_1 A_2 \cos(\delta_1 - \delta_2)\cos(\phi_1 - \phi_2)} . \quad (83)$$

The sign of strong phases δ_i is the same for A_{f+} and $\bar{A}_{f-}$ because CP is conserved by strong interactions. The weak phases have opposite signs.

The presence of hadronic uncertainties in A_i and of strong phases δ_i complicates the extraction of the phases ϕ_i from data. An example of this type of CP violation in K decays is ε'. We will demonstrate this below.

CP Violation in the Interference of Mixing and Decay. This type of CP violation is only possible in neutral B and K decays. We will use B decays for illustration suppressing the subscripts d and s. Moreover, we set $\Delta\Gamma = 0$. Formulae with $\Delta\Gamma \neq 0$ can be found in [9,16].

Most interesting are the decays into final states which are CP-eigenstates. Then a time dependent asymmetry defined by

$$a_{CP}(t, f) = \frac{\Gamma(B^0(t) \to f) - \Gamma(\bar{B}^0(t) \to f)}{\Gamma(B^0(t) \to f) + \Gamma(\bar{B}^0(t) \to f)} \tag{84}$$

is given by

$$a_{CP}(t, f) = a_{CP}^{\text{decay}}(f) \cos(\Delta M t) + a_{CP}^{\text{int}}(f) \sin(\Delta M t) \tag{85}$$

where we have separated the *decay* CP-violating contributions from those describing CP violation in the interference of mixing and decay:

$$a_{CP}^{\text{decay}}(f) = \frac{1 - |\xi_f|^2}{1 + |\xi_f|^2} \equiv C_f, \quad a_{CP}^{\text{int}}(f) = \frac{2\text{Im}\xi_f}{1 + |\xi_f|^2} \equiv -S_f . \tag{86}$$

Here C_f and S_f are popular notations found in the recent literature. The later type of CP violation is sometimes called the *mixing-induced* CP violation. The quantity ξ_f containing all the information needed to evaluate the asymmetries (86) is given by

$$\xi_f = \frac{q}{p} \frac{A(\bar{B}^0 \to f)}{A(B^0 \to f)} = \exp(i2\phi_M) \frac{A(\bar{B}^0 \to f)}{A(B^0 \to f)} \tag{87}$$

with ϕ_M, introduced in (74), denoting the weak phase in the $B^0 - \bar{B}^0$ mixing. $A(B^0 \to f)$ and $A(\bar{B}^0 \to f)$ are decay amplitudes. The time dependence of $a_{CP}(t, f)$ allows to extract a_{CP}^{decay} and a_{CP}^{int} as coefficients of $\cos(\Delta M t)$ and $\sin(\Delta M t)$, respectively.

Generally several decay mechanisms with different weak and strong phases can contribute to $A(B^0 \to f)$. These are tree diagram (current-current) contributions, QCD penguin contributions and electroweak penguin contributions. If they contribute with similar strength to a given decay amplitude the resulting CP asymmetries suffer from hadronic uncertainies related to matrix elements of the relevant operators Q_i. The situation is then analogous to the class just discussed. Indeed

$$\frac{A(\bar{B}^0 \to f)}{A(B^0 \to f)} = -\eta_f \left[\frac{A_T e^{i(\delta_T - \phi_T)} + A_P e^{i(\delta_P - \phi_P)}}{A_T e^{i(\delta_T + \phi_T)} + A_P e^{i(\delta_P + \phi_P)}} \right] \tag{88}$$

with $\eta_f = \pm 1$ being the CP-parity of the final state, depends on strong phases $\delta_{T,P}$ and hadronic matrix elements present in $A_{T,P}$. Thus the measurement of the asymmetry does not allow a clean determination of the weak phases $\phi_{T,P}$. The minus sign in (88) follows from our CP phase convention $CP|B^0\rangle = -|\bar{B}^0\rangle$, that has also been used in writing the phase factor in (87). Only ξ is phase convention independent. See Sect. 8.4.1 of [9] for details.

An interesting case arises when a single mechanism dominates the decay amplitude or the contributing mechanisms have the same weak phases. Then the hadronic matrix elements and strong phases drop out and

$$\frac{A(\bar{B}^0 \to f)}{A(B^0 \to f)} = -\eta_f e^{-i2\phi_D} \tag{89}$$

is a pure phase with ϕ_D being the weak phase in $A(B^0 \to f)$. Consequently

$$\xi_f = -\eta_f \exp(i2\phi_M) \exp(-i2\phi_D), \qquad |\xi_f|^2 = 1 . \tag{90}$$

In this particular case $a_{CP}^{\text{decay}}(f) = C_f$ vanishes and the CP asymmetry is given entirely in terms of the weak phases ϕ_M and ϕ_D:

$$a_{CP}(t, f) = \text{Im}\xi_f \sin(\Delta M t) \qquad \text{Im}\xi_f = \eta_f \sin(2\phi_D - 2\phi_M) = -S_f . \tag{91}$$

Thus the corresponding measurement of weak phases is free from hadronic uncertainties. A well known example is the decay $B_d \to \psi K_S$. Here $\phi_M = -\beta$ and $\phi_D = 0$. As in this case $\eta_f = -1$, we find

$$a_{CP}(t, f) = -\sin(2\beta)\sin(\Delta M t), \qquad S_f = \sin(2\beta) \tag{92}$$

which allows a very clean measurement of the angle β in the unitarity triangle. We will discuss other examples in Sect. 5.

We observe that the asymmetry $a_{CP}(t, f)$ measures directly the difference between the phases of $B^0 - \bar{B}^0$-mixing ($2\phi_M$) and of the decay amplitude ($2\phi_D$). This tells us immediately that we are dealing with the interference of mixing and decay. As ϕ_M and ϕ_D are phase convention dependent quantities, only their difference is physical, it is impossible to state on the basis of a single asymmetry whether CP violation takes place in the decay or in the mixing. To this end at least two asymmetries for $B^0(\bar{B}^0)$ decays to different final states f_i have to be measured. As ϕ_M does not depend on the final state, $\text{Im}\xi_{f_1} \neq \text{Im}\xi_{f_2}$ is a signal of CP violation in the decay.

We will see in Sect. 5 that the ideal situation presented above does not always take place and two or more different mechanism with different weak and strong phases contribute to the CP asymmetry. One finds then

$$a_{CP}(t, f) = C_f \cos(\Delta M t) - S_f \sin(\Delta M t), \tag{93}$$

$$C_f = -2r \sin(\phi_1 - \phi_2) \sin(\delta_1 - \delta_2) , \tag{94}$$

$$S_f = -\eta_f \left[\sin 2(\phi_1 - \phi_M) + 2r \cos 2(\phi_1 - \phi_M) \sin(\phi_1 - \phi_2) \cos(\delta_1 - \delta_2) \right] \tag{95}$$

where $r = A_2/A_1 \ll 1$ and ϕ_i and δ_i are weak and strong phases, respectively. For $r = 0$ the previous formulae are obtained.

In the case of K decays, this type of CP violation can be cleanly measured in the rare decay $K_L \to \pi^0 \nu \bar{\nu}$. Here the difference between the weak phase in the $K^0 - \bar{K}^0$ mixing and in the decay $\bar{s} \to \bar{d}\nu\bar{\nu}$ matters.

We can now compare the two classifications of different types of CP violation. CP violation in mixing is a manifestation of indirect CP violation. CP violation in decay is a manifestation of direct CP violation. CP violation in interference of mixing and decay contains elements of both the indirect and direct CP violation.

It is clear from this discussion that only in the case of the third type of CP violation there are possibilities to measure directly weak phases without hadronic uncertainties and moreover without invoking sophisticated methods. This takes place provided a single mechanism (diagram) is responsible for the decay or the contributing decay mechanisms have the same weak phases. However, we will see in Sect. 5 that there are other strategies, involving also decays to CP non-eigenstates, that provide clean measurements of the weak phases.

Another Look at ε and ε'. Let us finally investigate what type of CP violation is represented by ε and ε'. Here instead of different mechanisms it is sufficient to talk about different isospin amplitudes.

In the case of ε, CP violation in decay is not possible as only the isospin amplitude A_0 is involved. See (46). We know also that only $\mathrm{Re}\varepsilon = \mathrm{Re}\bar{\varepsilon}$ is related to CP violation in mixing. Consequently:

- $\mathrm{Re}\varepsilon$ represents CP violation in mixing,
- $\mathrm{Im}\varepsilon$ represents CP violation in the interference of mixing and decay.

In order to analyze the case of ε' we use the formula (49) to find

$$\mathrm{Re}\,\varepsilon' = -\frac{1}{\sqrt{2}} \left| \frac{A_2}{A_0} \right| \sin(\phi_2 - \phi_0) \sin(\delta_2 - \delta_0) \tag{96}$$

$$\mathrm{Im}\,\varepsilon' = \frac{1}{\sqrt{2}} \left| \frac{A_2}{A_0} \right| \sin(\phi_2 - \phi_0) \cos(\delta_2 - \delta_0) . \tag{97}$$

Consequently:

- Re ε' represents CP violation in decay as it is only non zero provided simultaneously $\phi_2 \neq \phi_0$ and $\delta_2 \neq \delta_0$.
- Im ε' exists even for $\delta_2 = \delta_0$ but as it requires $\phi_2 \neq \phi_0$ it represents CP violation in decay as well.

Experimentally $\delta_2 \neq \delta_0$. Within the SM, ϕ_2 and ϕ_0 are connected with electroweak penguins and QCD penguins, respectively. We will see in Sect. 4 that these phases differ from each other so that a nonvanishing ε' is obtained.

3   Standard Analysis of the Unitarity Triangle (UT)

3.1   General Procedure

After these general discussion of basic concepts let us concentrate on the standard analysis of the Unitarity Triangle (see Fig. 1) within the SM. A very detailed description of this analysis with the participation of the leading experimentalists and theorists in this field can be found in [3].

Setting $\lambda = |V_{us}| = 0.224$, the analysis proceeds in the following five steps:

Step 1:

From $b \to c$ transition in inclusive and exclusive leading B-meson decays one finds $|V_{cb}|$ and consequently the scale of the UT:

$$|V_{cb}| \quad \Longrightarrow \quad \lambda |V_{cb}| = \lambda^3 A . \tag{98}$$

Step 2:

From $b \to u$ transition in inclusive and exclusive B meson decays one finds $|V_{ub}/V_{cb}|$ and consequently using (12) the side $CA = R_b$ of the UT:

$$\left| \frac{V_{ub}}{V_{cb}} \right| \quad \Longrightarrow \quad R_b = \sqrt{\bar{\varrho}^2 + \bar{\eta}^2} = 4.35 \cdot \left| \frac{V_{ub}}{V_{cb}} \right| . \tag{99}$$

Step 3:

From the experimental value of ε_K in (65) and the formula (63) rewritten in terms of Wolfenstein parameters one derives the constraint on $(\bar{\varrho}, \bar{\eta})$ [38]

$$\bar{\eta} \left[(1 - \bar{\varrho}) A^2 \eta_2 S_0(x_t) + P_c(\varepsilon) \right] A^2 \hat{B}_K = 0.187, \tag{100}$$

where

$$P_c(\varepsilon) = [\eta_3 S_0(x_c, x_t) - \eta_1 x_c] \frac{1}{\lambda^4}, \qquad x_i = \frac{m_i^2}{M_W^2} \tag{101}$$

with all symbols defined in the previous section and $P_c(\varepsilon) = 0.29 \pm 0.07$ [36] summarizing the contributions of box diagrams with two charm quark exchanges and the mixed charm-top exchanges.

As seen in Fig. 5, equation (100) specifies a hyperbola in the $(\bar{\varrho}, \bar{\eta})$ plane. The position of the hyperbola depends on m_t, $|V_{cb}| = A\lambda^2$ and $\hat{B}_K$. With decreasing m_t, $|V_{cb}|$ and $\hat{B}_K$ it moves away from the origin of the $(\bar{\varrho}, \bar{\eta})$ plane.

Step 4:

From the measured ΔM_d and the formula (77), the side $AB = R_t$ of the UT can be determined:

$$R_t = \frac{1}{\lambda} \frac{|V_{td}|}{|V_{cb}|} = 0.85 \cdot \left[\frac{|V_{td}|}{7.8 \cdot 10^{-3}} \right] \left[\frac{0.041}{|V_{cb}|} \right], \tag{102}$$

$$|V_{td}| = 7.8 \cdot 10^{-3} \left[\frac{230\,\text{MeV}}{\sqrt{\hat{B}_{B_d}} F_{B_d}} \right] \left[\frac{167\,GeV}{\overline{m}_t(m_t)} \right]^{0.76} \left[\frac{\Delta M_d}{0.50/\text{ps}} \right]^{0.5} \sqrt{\frac{0.55}{\eta_B}} \tag{103}$$

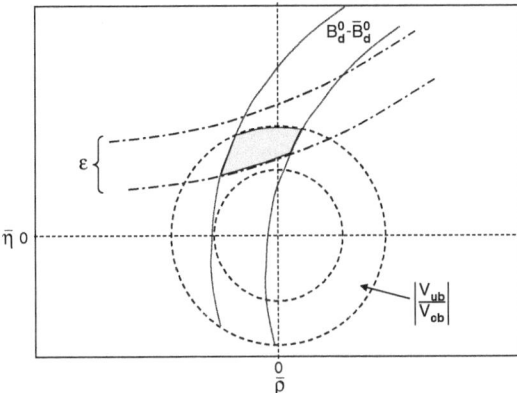

Fig. 5. Schematic determination of the Unitarity Triangle.

with all symbols defined in the previous section. $\overline{m}_t(m_t) = (167 \pm 5)$ GeV. Note that R_t suffers from additional uncertainty in $|V_{cb}|$, which is absent in the determination of $|V_{td}|$ this way. The constraint in the $(\bar{\varrho}, \bar{\eta})$ plane coming from this step is illustrated in Fig. 5.

Step 5:

The measurement of ΔM_s together with ΔM_d allows to determine R_t in a different manner:

$$R_t = 0.90 \left[\frac{\xi}{1.24}\right] \sqrt{\frac{18.4/ps}{\Delta M_s}} \sqrt{\frac{\Delta M_d}{0.50/ps}}, \qquad \xi = \frac{\sqrt{\hat{B}_{B_s}} F_{B_s}}{\sqrt{\hat{B}_{B_d}} F_{B_d}}. \qquad (104)$$

One should note that m_t and $|V_{cb}|$ dependences have been eliminated this way and that ξ should in principle contain much smaller theoretical uncertainties than the hadronic matrix elements in ΔM_d and ΔM_s separately.

The main uncertainties in these steps originate in the theoretical uncertainties in $\hat{B}_K$ and $\sqrt{\hat{B}_d} F_{B_d}$ and to a lesser extent in ξ [3]:

$$\hat{B}_K = 0.86 \pm 0.15, \quad \sqrt{\hat{B}_d} F_{B_d} - (235^{+33}_{-41})\ MeV, \quad \xi - 1.24 \pm 0.08 \qquad (105)$$

Also the uncertainties due to $|V_{ub}/V_{cb}|$ in step 2 are substantial. The QCD sum rules results for the parameters in question are similar and can be found in [3]. Finally [3]

$$\Delta M_d = (0.503 \pm 0.006)/ps, \qquad \Delta M_s > 14.4/ps \ \text{at}\ 95\% \ \text{C.L.} \qquad (106)$$

3.2   The Angle β from $B_d \to \psi K_S$

One of the highlights of the year 2002 were the considerably improved measurements of $\sin 2\beta$ by means of the time-dependent CP asymmetry

$$a_{\psi K_S}(t) \equiv -a_{\psi K_S} \sin(\Delta M_d t) = -\sin 2\beta \sin(\Delta M_d t) . \qquad (107)$$

The BaBar [39] and Belle [40] collaborations find

$$(\sin 2\beta)_{\psi K_S} = \begin{cases} 0.741 \pm 0.067\,(\text{stat}) \pm 0.033\,(\text{syst})\ (\text{BaBar}) \\ 0.719 \pm 0.074\,(\text{stat}) \pm 0.035\,(\text{syst})\ (\text{Belle}). \end{cases}$$

Combining these results with earlier measurements by CDF $(0.79^{+0.41}_{-0.44})$, ALEPH $(0.84^{+0.82}_{-1.04} \pm 0.16)$ and OPAL gives the grand average [41]

$$(\sin 2\beta)_{\psi K_S} = 0.734 \pm 0.054 \ . \tag{108}$$

This is a milestone in the field of CP violation and in the tests of the SM as we will see in a moment. Not only violation of this symmetry has been confidently established in the B system, but also its size has been measured very accurately. Moreover in contrast to the five constraints listed above, the determination of the angle β in this manner is theoretically very clean.

3.3   Unitarity Triangle 2003

We are now in the position to combine all these constraints in order to construct the unitarity triangle and determine various quantities of interest. In this context the important issue is the error analysis of these formulae, in particular the treatment of theoretical uncertainties. In the literature the most popular are the Bayesian approach [42] and the frequentist approach [43]. For the PDG analysis see [19]. A critical comparison of these and other methods can be found in [3]. I can recommend this reading.

In Fig. 6 we show the result of the recent update of an analysis in collaboration with Parodi and Stocchi [44] that uses the Bayesian approach. The results presented below are very close to the ones presented in [3] that was led by my Italian collaborators. The allowed region for $(\bar\varrho, \bar\eta)$ is the area inside the smaller ellipse. We observe that the region $\bar\varrho < 0$ is disfavoured by the lower bound on ΔM_s. It is clear from this figure that the measurement of ΔM_s giving R_t through (104) will have a large impact on the plot in Fig. 6.

The ranges for various quantities that result from this analysis are given in the SM column of Table 1. The UUT column will be discussed in Sect. 7. The SM results follow from the five steps listed above and (108) implying an impressive precision on the angle β:

$$(\sin 2\beta)_{\text{tot}} = 0.705^{+0.042}_{-0.032}, \qquad \beta = (22.4 \pm 1.5)^\circ \ . \tag{109}$$

On the other hand $(\sin 2\beta)_{\text{ind}}$ obtained by means of the five steps only is found to be [44]

$$(\sin 2\beta)_{ind} = 0.685 \pm 0.052 \tag{110}$$

demonstrating an excellent agreement (see also Fig. 6) between the direct measurement in (108) and the standard analysis of the UT within the SM. This gives a strong indication that the CKM matrix is very likely the dominant source of CP violation in flavour violating decays. In order to be sure

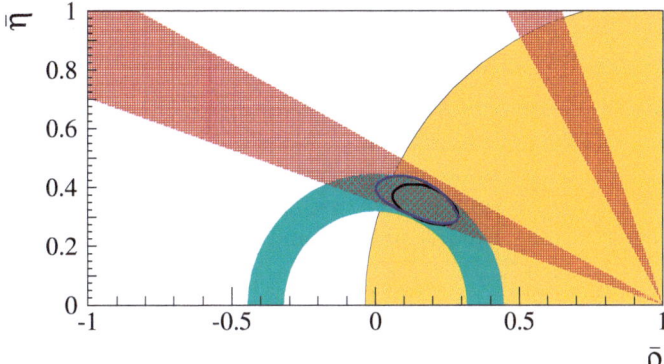

Fig. 6. The allowed 95% regions in the $(\bar{\varrho}, \bar{\eta})$ plane in the SM (narrower region) and in the MFV models (broader region) from the update of [44]. The individual 95% regions for the constraint from $\sin 2\beta$, ΔM_s and R_b are also shown.

whether this is indeed the case other theoretically clean quantities have to be measured. In particular the angle γ that is more sensitive to new physics contributions than β. We will return to other processes that are useful for the determination of the UT in Sects. 5 and 6.

Table 1. Values for different quantities from the update of [44]. $\lambda_t = V_{ts}^* V_{td}$.

Strategy	UUT	SM
$\bar{\eta}$	0.361 ± 0.032	0.341 ± 0.028
$\bar{\varrho}$	0.149 ± 0.056	0.178 ± 0.046
$\sin 2\beta$	$0.715^{+0.037}_{-0.034}$	$0.705^{+0.042}_{-0.032}$
$\sin 2\alpha$	0.03 ± 0.31	-0.19 ± 0.25
γ	$(67.5 \pm 8.9)^\circ$	$(61.5 \pm 7.0)^\circ$
R_b	0.393 ± 0.025	0.390 ± 0.024
R_t	0.925 ± 0.060	0.890 ± 0.048
ΔM_s (ps^{-1})	$17.3^{+2.1}_{-1.3}$	$18.3^{+1.7}_{-1.5}$
$|V_{td}|$ (10^{-3})	8.61 ± 0.55	8.24 ± 0.41
$\mathrm{Im}\lambda_t$ (10^{-4})	1.39 ± 0.12	1.31 ± 0.10

4 ε'/ε in the Standard Model

4.1   Preliminaries

The ratio ε'/ε that parametrizes the size of direct CP violation with respect to the indirect CP violation in $K_L \to \pi\pi$ decays has been the subject of very intensive experimental and theoretical studies in the last three decades. After tremendous efforts, on the experimental side the world average based on the recent results from NA48 [45] and KTeV [46], and previous results from NA31 [47] and E731 [48], reads

$$\varepsilon'/\varepsilon = (16.6 \pm 1.6) \cdot 10^{-4} \qquad (2003) . \qquad (111)$$

On the other hand, the theoretical estimates of this ratio are subject to very large hadronic uncertainties. While several analyzes of recent years within the Standard Model (SM) find results that are compatible with (111) [49–56]), it is fair to say that the chapter on the theoretical calculations of ε'/ε is far from being closed. A full historical account of the theoretical efforts before 1998 can for example be found in [4,57]. See also [58].

It should be emphasized that all existing analyzes of ε'/ε use the NLO Wilson coefficients calculated by the Munich and Rome groups in 1993 [59–61] but the hadronic matrix elements, the main theoretical uncertainty in ε'/ε, vary from paper to paper. Nevertheless, apart from the hadronic matrix element of the dominant QCD penguin operator Q_6, in the last years progress has been made with the determination of all other relevant parameters, which enter the theoretical prediction of ε'/ε. Let me review then briefly the present situation. Further details can be found in [62].

4.2   Basic Formulae

The central formula for ε'/ε of [49,50,60,62] can be cast into the following approximate expression (it reproduces the results in [62] to better than 2%)

$$\frac{\varepsilon'}{\varepsilon} = \mathrm{Im}\lambda_t \cdot F_{\varepsilon'}(x_t), \qquad \lambda_t = V_{ts}^* V_{td}, \qquad (112)$$

$$F_{\varepsilon'}(x_t) = [18.7\, R_6(1 - \Omega_{\mathrm{IB}}) - 6.9\, R_8 - 1.8] \left[\frac{\Lambda_{\overline{\mathrm{MS}}}^{(4)}}{340\,\mathrm{MeV}} \right] \qquad (113)$$

with the non-perturbative parameters R_6 and R_8 defined as

$$R_6 \equiv B_6^{(1/2)} \left[\frac{121\,\mathrm{MeV}}{m_s(m_c) + m_d(m_c)} \right]^2 , \qquad R_8 \equiv \frac{B_8^{(3/2)}}{B_6^{(1/2)}} R_6. \qquad (114)$$

The hadronic B-parameters $B_6^{(1/2)}$ and $B_8^{(3/2)}$ represent the matrix elements of the dominant QCD-penguin (Q_6) and the dominant electroweak penguin

(Q_8) operator. In the large–N_c approach of [63] they are given by [60,64]

$$\langle Q_6 \rangle_0 = -4\sqrt{\frac{3}{2}}\,(F_K - F_\pi)\left(\frac{m_K^2}{121\,\text{MeV}}\right)^2 R_6 = -0.597 \cdot R_6 \,\text{GeV}^3\,, \quad (115)$$

$$\langle Q_8 \rangle_2 = \sqrt{3}\,F_\pi\left(\frac{m_K^2}{121\,\text{MeV}}\right)^2 R_8 = 0.948 \cdot R_8 \,\text{GeV}^3\,. \quad (116)$$

Finally $\Omega_{\text{IB}} = 0.06 \pm 0.08$ [65] represents the isospin breaking correction.

In the strict large–N_c limit, $B_6^{(1/2)} = B_8^{(3/2)} = 1$ and

$$\frac{R_6}{R_8} = 1\,, \qquad \frac{\langle Q_6 \rangle_0}{\langle Q_8 \rangle_2} = -0.63\,, \quad (117)$$

so that there is a one-to-one correspondence between ε'/ε and $R_6 = R_8$ for fixed values of the remaining parameters. Moreover, only for certain values of $m_s(m_c)$ is one able to obtain the experimental value for ε'/ε [66]. Note that once $m_s(m_c)$ is known, also R_6, R_8, $\langle Q_6 \rangle_0$ and $\langle Q_8 \rangle_2$ are known, but they always satisfy the relations in (117).

4.3   Numerical Results

The relevant input parameters are as follows. First

$$\text{Im}\lambda_t = (1.31 \pm 0.10) \cdot 10^{-4}, \qquad m_s(m_c) = 115 \pm 20 \,\text{MeV} \quad (118)$$

where the value of m_s is an average over recent determinations (see references in [62]). The central value corresponds to $m_s(2\,\text{GeV}) = 100$ MeV.

Concerning $\langle Q_8 \rangle_2$, in the last years progress has been achieved both in the framework of lattice QCD as well as with analytic methods [55,56,67–70]. The current status of $\langle Q_8 \rangle_2$ has been summarized nicely in [69]. The most precise determination of $\langle Q_8 \rangle_2$ comes from the lattice QCD measurement [68], corresponding to $R_8 = 1.0 \pm 0.1$. Several analytic methods give higher results but compatible with it. We will use [62]

$$R_8 = 1.0 \pm 0.2\,, \qquad \langle Q_8 \rangle_2^{\text{NDR}}(m_c) = (0.95 \pm 0.19)\,\text{GeV}^3\,. \quad (119)$$

The situation is less clear concerning $\langle Q_6 \rangle_0$ but assuming that new physics contributions to ε'/ε can be neglected one finds from (111)–(113) and (119) that [62]

$$R_6 = 1.23 \pm 0.16\,, \qquad \langle Q_6 \rangle_0^{\text{NDR}}(m_c) = -(0.73 \pm 0.10)\,\text{GeV}^3\,. \quad (120)$$

More generally, the correlation between R_8 and R_6 that is implied by the data on ε'/ε is shown in the spirit of the "ε'/ε–path" of [71] (see also [72,56]) in Fig. 7 [62]. The solid straight line corresponds to the central values of parameters, whereas the short-dashed lines are the uncertainties due to a

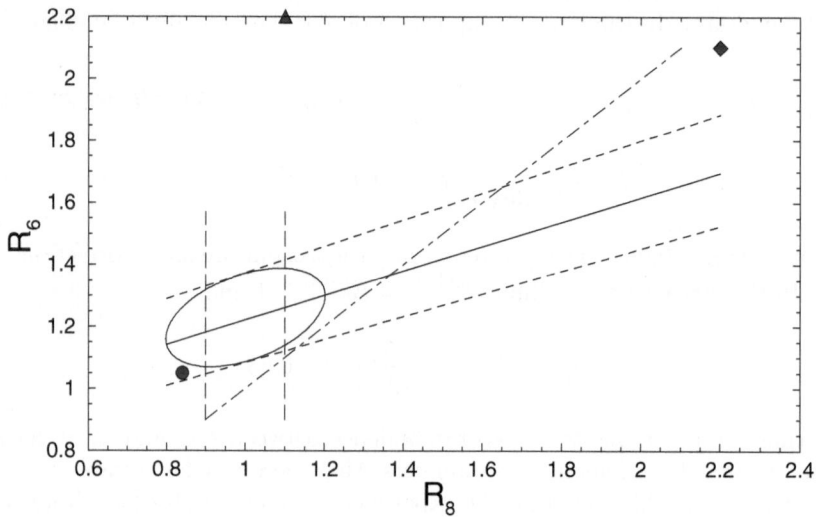

Fig. 7. R_6 as a function of R_8. For a detailed explanation see the text.

variation of the input parameters. The vertical long-dashed lines indicate the lattice range for R_8 [68], whereas the ellipse describes the correlation between R_6 and R_8 implied by the data on ε'/ε when taking into account the more conservative constraint on R_8 given in (119). The value in (120) corresponds to this ellipse. The full circle, triangle and diamond in Fig. 7 represent the central results of [54] [55] and [56] respectively, that are discussed in detail in [62], and the dashed-dotted line shows the strict large-N_c relation $R_6 = R_8$.

In Table 2 we show ε'/ε for specific values of R_6, R_8 and $\Lambda_{\overline{\text{MS}}}^{(4)}$, and in Table 3 as a function of $m_s(m_c)$ and $\Lambda_{\overline{\text{MS}}}^{(4)}$ obtained in the strict large–N_c limit. Here $\text{Im}\lambda_t = 1.34 \cdot 10^{-4}$.

4.4   Conclusions

There are essentially two messages from this analysis [62]:

- If indeed $R_8 = 1.0 \pm 0.2$ as indicated by several recent estimates, then the data on ε'/ε imply

$$R_6 = 1.23 \pm 0.16 , \qquad \frac{R_6}{R_8} \approx 1.23 \pm 0.29 , \qquad (121)$$

$$\langle Q_6 \rangle_0^{\text{NDR}}(m_c) = -(0.77 \pm 0.19) \langle Q_8 \rangle_2^{\text{NDR}}(m_c) . \qquad (122)$$

This is in accordance with the results in [54], but due to significant uncertainties is also compatible with the large–N_c approach in [63] in which

Table 2. ε'/ε in units of 10^{-4} for $m_t = 165\,\text{GeV}$ and various $\Lambda_{\overline{\text{MS}}}^{(4)}$, R_6, R_8.

$\Lambda_{\overline{\text{MS}}}^{(4)}$ [MeV]	$R_6 = 1.00$		$R_6 = 1.20$		$R_6 = 1.40$	
	$R_8 = 0.8$	$R_8 = 1.0$	$R_8 = 0.8$	$R_8 = 1.0$	$R_8 = 0.8$	$R_8 = 1.0$
310	12.2	10.5	16.4	14.8	20.7	19.0
340	13.3	11.5	17.9	16.1	22.5	20.7
370	14.6	12.6	19.5	17.6	24.5	22.5

Table 3. The ratio ε'/ε in units of 10^{-4} for the strict large–N_c results $B_6^{(1/2)} = B_8^{(3/2)} = 1.0$, $\hat{B}_K = 0.75$ and various values of $\Lambda_{\overline{\text{MS}}}^{(4)}$ and $m_s(m_c)$.

$\Lambda_{\overline{\text{MS}}}^{(4)}$ [MeV]	$m_s(m_c) = 115\,\text{MeV}$	$m_s(m_c) = 105\,\text{MeV}$	$m_s(m_c) = 95\,\text{MeV}$
310	10.8	13.4	16.8
340	11.8	14.6	18.2
370	12.9	15.9	19.9

$R_6 \approx R_8$. However the strong suppression of $\langle Q_6 \rangle_0^{\text{NDR}}(m_c)$ relatively to $\langle Q_8 \rangle_2^{\text{NDR}}(m_c)$ is not fully supported by the data.

- The large–N_c approach of [63] can only be made consistent with data provided

$$R_6 = R_8 = 1.36 \pm 0.30 \tag{123}$$

and $\langle Q_8 \rangle_2^{\text{NDR}}(m_c)$ is higher than obtained by most recent approaches reviewed in [62,69]. This requires $m_s(m_c) \leq 105\,\text{MeV}$ which is on the low side of (118) but close to low values of $m_s(m_c)$ indicated by the most recent lattice simulations with dynamical fermions [73,74].

Large non-factorizable contributions to $\langle Q_6 \rangle_0$ and $\langle Q_8 \rangle_2$ are found in the chiral limit in the large N approach of [56] and consequently the structure of the matrix elements in question differs in this approach from the formulae (115). Interestingly, in spite of these large non-factorizable contributions, the relation $R_6 = R_8$ is roughly satisfied in this approach (see Fig. 7) but one has to go beyond the chiral limit to draw definite conclusions.

As seen in Fig. 7, all these three scenarios are consistent with the data. Which of these pictures of ε'/ε is correct, can only be answered by calculating $\langle Q_6 \rangle_0$, $\langle Q_8 \rangle_2$ and m_s accurately by means of non-perturbative methods that are reliable. Such calculations are independent of the assumption about the role of new physics in ε'/ε that has been made in [62] in order to extract $\langle Q_6 \rangle_0$ from the data. If the values for $R_{6,8}$ will be found one day to lie significantly outside the allowed region in Fig. 7, new physics contributions to ε'/ε will be required in order to fit the experimental data.

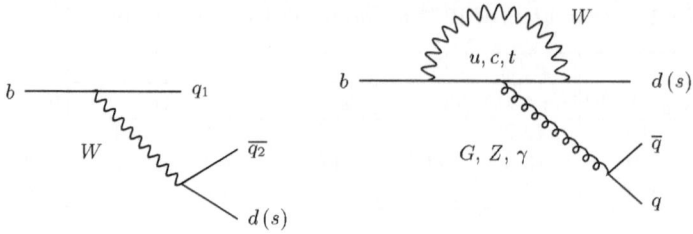

Fig. 8. Tree and penguin diagrams.

5   The Angles α, β and γ from B Decays

5.1   Preliminaries

CP violation in B decays is certainly one of the most important targets of B-factories and of dedicated B-experiments at hadron facilities. It is well known that CP-violating effects are expected to occur in a large number of channels at a level attainable experimentally in the near future. Moreover there exist channels which offer the determination of CKM phases essentially without any hadronic uncertainties.

The first results on $\sin 2\beta$ from BaBar and Belle are very encouraging. These results should be further improved over the coming years through the new measurements of $a_{\psi K_S}(t)$ by both collaborations and by CDF and D0 at Fermilab. Moreover measurements of CP asymmetries in other B decays and the measurements of the angles α, β and γ by means of various strategies using two-body B decays should contribute substantially to our understanding of CP violation and will test the KM picture of CP violation.

The various types of CP violation have been already classified in Sect. 2. It turned out that CP violation in the interference of mixing and decay, in a B meson decay into a CP eigenstate, is very suitable for a theoretically clean determination of the angles of the unitarity triangle provided a single CKM phase governs the decay. However as we will see below several useful strategies for the determination of the angles α, β and γ have been developed that are effective also in the presence of competing CKM phases and when the final state in not a CP eigenstate. The notes below should only be considered as an introduction to this rich field. For more details the references in Sect. 1 should be contacted.

5.2   Classification of Elementary Processes

Non-leptonic B decays are caused by elementary decays of b quarks that are represented by tree and penguin diagrams in Fig. 8. Generally we have

$$b \to q_1 \bar{q}_2 d(s), \qquad b \to q \bar{q} d(s) \tag{124}$$

for tree and penguin diagrams, respectively.

There are twelve basic transitions that can be divided into three classes:

Class I: both tree and penguin diagrams contribute. Here $q_1 = q_2 = q = u, c$ and consequently the basic transitions are

$$b \to c\bar{c}s, \qquad b \to c\bar{c}d, \qquad b \to u\bar{u}s, \qquad b \to u\bar{u}d. \qquad (125)$$

Class II: only tree diagrams contribute. Here $q_1 \neq q_2 \in \{u, c\}$ and

$$b \to c\bar{u}s, \qquad b \to c\bar{u}d, \qquad b \to u\bar{c}s, \qquad b \to u\bar{c}d. \qquad (126)$$

Class III: only penguin diagrams contribute. Here $q = d, s$ and

$$b \to s\bar{s}s, \qquad b \to s\bar{s}d, \qquad b \to d\bar{d}s, \qquad b \to d\bar{d}d. \qquad (127)$$

Now in presenting various decays below, we did not show the corresponding diagrams on purpose. Afterall these are lectures and the exercise for the students is to draw these diagrams by embedding the elementary diagrams of Fig. 8 into a given B meson decay. In case of difficulties the student should look at [14,16] where these diagrams can be found.

5.3 Neutral B Decays into CP Eigenstates

$B_d^0 \to J/\psi K_S$ **and** β. The amplitude for this decay can be written as follows

$$A(B_d^0 \to J/\psi K_S) = V_{cs} V_{cb}^* (A_T + P_c) + V_{us} V_{ub}^* P_u + V_{ts} V_{tb}^* P_t \qquad (128)$$

where A_T denotes tree diagram contributions and P_i with $i = u, c, t$ stand for penguin diagram contributions with internal u, c and t quarks. Now

$$V_{cs} V_{cb}^* \approx A\lambda^2, \quad V_{us} V_{ub}^* \approx A\lambda^4 R_b e^{i\gamma}, \quad V_{ts} V_{tb}^* = -V_{us} V_{ub}^* - V_{cs} V_{cb}^* \qquad (129)$$

with the last relation following from the unitarity of the CKM matrix. Thus

$$A(B_d^0 \to J/\psi K_S) = V_{cs} V_{cb}^* (A_T + P_c - P_t) + V_{us} V_{ub}^* (P_u - P_t) . \qquad (130)$$

We next note that

$$\left| \frac{V_{us} V_{ub}^*}{V_{cs} V_{cb}^*} \right| \leq 0.02, \qquad \frac{P_u - P_t}{A_T + P_c} \frac{}{P_t} \ll 1 \qquad (131)$$

where the last inequality is very plausible as the Wilson coefficients of the current–current operators responsible for A_T are much larger than the ones of the penguin operators [4,8]. Consequently this decay is dominated by a single CKM factor and as discussed in Sect. 2, a clean determination of the relevant CKM phase is possible. Indeed in this decay $\phi_D = 0$ and $\phi_M = -\beta$. Using (91) we find then ($\eta_{J/\psi K_S} = -1$)

$$a_{CP}^{\text{int}}(J/\psi K_S) = \eta_{J/\psi K_S} \sin(2\phi_D - 2\phi_M) = -\sin 2\beta, \qquad (132)$$

$$C_{J/\psi K_S} = 0, \qquad S_{J/\psi K_S} = \sin 2\beta \qquad (133)$$

that is confirmed by experiment as discussed in Sect. 3.

$B_s^0 \to J/\psi\phi$ and β_s. This decay differs from the previous one by the spectator quark, with $d \to s$ so that the formulae above remain unchanged except that now $\phi_M = -\beta_s = -\lambda^2\bar{\eta}$. A complication arises as the final state is an admixture of $CP = +$ and $CP = -$ states. This issue can be resolved experimentally [14]. Choosing $\eta_{J/\psi\phi} = 1$ we then find

$$a_{CP}^{int}(J/\psi\phi) = \sin(2\phi_D - 2\phi_M) = 2\beta_s = 2\lambda^2\bar{\eta} \approx 0.03, \qquad C_{J/\psi\phi} = 0 . \quad (134)$$

Thus this asymmetry measures the phase of V_{ts} that is predicted to be very small from the unitarity of the CKM matrix alone. Because of this there is a lot of room for new physics contributions here.

$B_d^0 \to \phi K_S$ and β. This decay proceeds entirely through penguin diagrams and consequently should be much more sensitive to new physics contributions than the decay $B_d^0 \to J/\psi K_S$. Assuming $\phi = (s\bar{s})$, the decay amplitude is given by (130) with A_T removed:

$$A(B_d^0 \to \phi K_S) = V_{cs}V_{cb}^*(P_c - P_t) + V_{us}V_{ub}^*(P_u - P_t) . \quad (135)$$

With

$$\left| \frac{V_{us}V_{ub}^*}{V_{cs}V_{cb}^*} \right| \leq 0.02, \qquad \frac{P_u - P_t}{P_c - P_t} = \mathcal{O}(1) \quad (136)$$

also in this decay single CKM phase dominates and as ϕ_D and ϕ_M are the same as in $B_d^0 \to J/\psi K_S$ we find

$$C_{\phi K_S} = 0, \qquad S_{\phi K_S} = S_{J/\psi K_S} = \sin 2\beta . \quad (137)$$

The equality of these two asymmetries need not be perfect as the ϕ meson is not entirely a $s\bar{s}$ state and the approximation of neglecting the second amplitude in (135) could be only true within a few percent. However, a detailed analysis shows [75] that these two asymmetries should be very close to each other within the SM: $|S_{\phi K_S} - S_{J/\psi K_S}| \leq 0.04$. Any strong violation of this bound would be a signal for new physics.

In view of this prediction, the first results on this asymmetry from BaBar [76] and Belle [77] are truely exciting:

$$(\sin 2\beta)_{\phi K_S} = \begin{cases} -0.19 \pm 0.51 \,(\text{stat}) \pm 0.09 \,(\text{syst}) \,(\text{BaBar}) \\ -0.73 \pm 0.64 \,(\text{stat}) \pm 0.18 \,(\text{syst}) \,(\text{Belle}). \end{cases}$$

Consequently

$$S_{\phi K_s} = -0.39 \pm 0.41, \qquad C_{\phi K_s} = 0.56 \pm 0.43, \quad (138)$$

$$|S_{\phi K_S} - S_{J/\psi K_S}| = 1.12 \pm 0.41 \quad (139)$$

where the result for $C_{\phi K_S}$, that is consistent with zero, comes solely from Belle. We observe that the bound $|S_{\phi K_S} - S_{J/\psi K_S}| \leq 0.04$ is violated by

2.7σ. While this is still insufficient to claim the presence of new physics, the fact that the two asymmetries are found to be quite different, invited a number of theorists to speculate what kind of new physics could be responsible for this difference. Some references are given in [78]. Enhanced QCD penguins, enhanced Z^0 penguins, rather involved supersymmetric scenarios have been suggested as possible origins of the departure from the SM prediction. I have no space to review these papers and although I find a few of them quite interesting, it is probably better to wait until the experimental errors decrease.

Of interest are also the measurements

$$S_{\eta' K_S} = \begin{cases} 0.02 \pm 0.35 \text{ (BaBar)} \\ 0.76 \pm 0.36 \text{ (Belle)}. \end{cases}$$

and $C_{\eta' K_S} = -0.26 \pm 0.22$ from Belle that are fully consistent with $(\sin 2\beta)_{J/\psi K_S}$. At first sight one could wonder why this asymmetry differs from $(\sin 2\beta)_{\phi K_S}$ as the decay in question is also penguin dominated and $\eta' \approx (s\bar{s})$, but the fact that η' deviates from a pure $(s\bar{s})$ state more than ϕ allows for some small contributions involving tree diagrams that could spoil the exact equality of these two asymmetries.

$B_d^0 \to \pi^+ \pi^-$ and α. This decay receives the contributions from both tree and penguin diagrams. The amplitude can be written as follows

$$A(B_d^0 \to \pi^+ \pi^-) = V_{ud} V_{ub}^* (A_T + P_u) + V_{cd} V_{cb}^* P_c + V_{td} V_{tb}^* P_t \tag{140}$$

where

$$V_{cd} V_{cb}^* \approx A\lambda^3, \ \ V_{ud} V_{ub}^* \approx A\lambda^3 R_b e^{i\gamma}, \ \ V_{td} V_{tb}^* = -V_{ud} V_{ub}^* - V_{cd} V_{cb}^* . \tag{141}$$

Consequently

$$A(B_d^0 \to \pi^+ \pi^-) = V_{ud} V_{ub}^* (A_T + P_u - P_t) + V_{cd} V_{cb}^* (P_c - P_t). \tag{142}$$

We next note that

$$\left| \frac{V_{cd} V_{cb}^*}{V_{ud} V_{ub}^*} \right| = \frac{1}{R_b} \approx 2.5, \qquad \frac{P_c - P_t}{A_T + P_u - P_t} \equiv \frac{P_{\pi\pi}}{T_{\pi\pi}}. \tag{143}$$

Now the dominance of a single CKM amplitude in contrast to the cases considered until now is very uncertain and takes only place provided $P_{\pi\pi} \ll T_{\pi\pi}$. Let us assume first that this is indeed the case. Then this decay is dominated by a single CKM factor and a clean determination of the relevant CKM phase is possible. Indeed in this decay $\phi_D = \gamma$ and $\phi_M = -\beta$. Using (91) we find then $(\eta_{\pi\pi} = 1)$

$$a_{CP}^{\text{int}}(\pi\pi) = \eta_{\pi\pi} \sin(2\phi_D - 2\phi_M) = \sin 2(\gamma + \beta) = -\sin 2\alpha \tag{144}$$

and

$$C_{\pi\pi} = 0, \qquad S_{\pi\pi} = \sin 2\alpha . \qquad (145)$$

This should be compared with the first results from BaBar and Belle:

$$C_{\pi\pi} = \begin{cases} -0.30 \pm 0.25 \,(\text{stat}) \pm 0.04 \,(\text{syst}) \,(\text{BaBar}) \\ -0.77 \pm 0.27 \,(\text{stat}) \pm 0.08 \,(\text{syst}) \,(\text{Belle}) \end{cases}$$

$$S_{\pi\pi} = \begin{cases} -0.02 \pm 0.34 \,(\text{stat}) \pm 0.05 \,(\text{syst}) \,(\text{BaBar}) \\ -1.23 \pm 0.41 \,(\text{stat}) \pm 0.08 \,(\text{syst}) \,(\text{Belle}). \end{cases}$$

The results from BaBar are consistent with our expectations. Afterall α from the UT fit is in the ballpark of 90°. On the other hand Belle results indicate a non-zero asymmetry and a sizable contribution of the penguin diagrams invalidating our assumption $P_{\pi\pi} \ll T_{\pi\pi}$. Yet, as the results from BaBar and Belle are incompatible with each other, the present picture of this decay is not conclusive and one has to wait for better data.

The "QCD penguin pollution" discussed above has to be taken care of in order to extract α. The well known strategy to deal with this "penguin problem" is the isospin analysis of Gronau and London [79]. It requires however the measurement of $Br(B^0 \to \pi^0\pi^0)$ which is expected to be below 10^{-6}: a very difficult experimental task. For this reason several, rather involved, strategies have been proposed which avoid the use of $B_d \to \pi^0\pi^0$ in conjunction with $a_{CP}(\pi^+\pi^-, t)$. They are reviewed in [9,13,14,16]. The most recent analyses of $B \to \pi\pi$, also related to the determination of γ and $(\bar{\varrho}, \bar{\eta})$, can be found in [3,80,81].

While I have some doubts that a precise value of α will follow in a foreseable future from this enterprise, one should also stress [82,83,44] that only a moderately precise measurement of $\sin 2\alpha$ can be as useful for the UT as a precise measurement of the angle β. This is clear from Table 1 that shows very large uncertainties in the indirect determination of $\sin 2\alpha$.

5.4   Decays to CP Non-eigenstates

Preliminaries. The strategies discussed below have the following general properties:

- $B_d^0(B_s^0)$ and their antiparticles $\bar{B}_d^0(\bar{B}_s^0)$ can decay to the same final state,
- Only tree diagrams contribute to the decay amplitudes,
- A full time dependent analysis of the four processes is required:

$$B_{d,s}^0(t) \to f, \quad \bar{B}_{d,s}^0(t) \to f, \quad B_{d,s}^0(t) \to \bar{f}, \quad \bar{B}_{d,s}^0(t) \to \bar{f} . \qquad (146)$$

The latter analysis allows to measure

$$\xi_f = \exp(i2\phi_M)\frac{A(\bar{B}^0 \to f)}{A(B^0 \to f)}, \qquad \xi_{\bar{f}} = \exp(i2\phi_M)\frac{A(\bar{B}^0 \to \bar{f})}{A(B^0 \to \bar{f})}. \qquad (147)$$

It turns out then that

$$\xi_f \cdot \xi_{\bar{f}} = F(\gamma, \phi_M) \tag{148}$$

without any hadronic uncertainties, so that determining ϕ_M from other decays as discussed above, allows the determination of γ. Let us show this.

$B_d^0 \to D^\pm \pi^\mp$, $\bar{B}_d^0 \to D^\pm \pi^\mp$ and γ. With $f = D^+ \pi^-$ the four decay amplitudes are given by

$$A(B_d^0 \to D^+ \pi^-) = M_f A \lambda^4 R_b e^{i\gamma}, \quad A(\bar{B}_d^0 \to D^+ \pi^-) = \bar{M}_f A \lambda^2 \tag{149}$$

$$A(\bar{B}_d^0 \to D^- \pi^+) = \bar{M}_{\bar{f}} A \lambda^4 R_b e^{-i\gamma}, \quad A(B_d^0 \to D^- \pi^+) = M_{\bar{f}} A \lambda^2 \tag{150}$$

where we have factored out the CKM parameters, A is a Wolfenstein parameter and M_i stand for the rest of the amplitudes that generally are subject to large hadronic uncertainties. The important point is that each of these transitions receives the contribution from a single phase so that

$$\xi_f^{(d)} = e^{-i(2\beta+\gamma)} \frac{1}{\lambda^2 R_b} \frac{\bar{M}_f}{M_f}, \quad \xi_{\bar{f}}^{(d)} = e^{-i(2\beta+\gamma)} \lambda^2 R_b \frac{\bar{M}_{\bar{f}}}{M_{\bar{f}}}. \tag{151}$$

Now, as CP is conserved in QCD we simply have

$$M_f = \bar{M}_{\bar{f}}, \quad \bar{M}_f = M_{\bar{f}} \tag{152}$$

and consequently [84]

$$\xi_f^{(d)} \cdot \xi_{\bar{f}}^{(d)} = e^{-i2(2\beta+\gamma)} \tag{153}$$

as promised. The phase β is already known with high precision and consequently γ can be determined. Unfortunately as seen in (149) and (150), the relevant intereferences are $\mathcal{O}(\lambda^2)$ and the execution of this strategy is a very difficult experimental task. See [85] for an interesting discussion.

$B_s^0 \to D_s^\pm K^\mp$, $\bar{B}_s^0 \to D_s^\pm K^\mp$ and γ. Replacing the d-quark by the s-quark in the strategy just discussed allows to solve the latter problem. With $f = D_s^+ K^-$ equations (149) and (150) are replaced by

$$A(B_s^0 \to D_s^+ K^-) = M_f A \lambda^3 R_b e^{i\gamma}, \quad A(\bar{B}_s^0 \to D_s^+ K^-) = \bar{M}_f A \lambda^3 \tag{154}$$

$$A(\bar{B}_s^0 \to D_s^- K^+) = \bar{M}_{\bar{f}} A \lambda^3 R_b e^{-i\gamma}, \quad A(B_s^0 \to D_s^- K^+) = M_{\bar{f}} A \lambda^3. \tag{155}$$

Proceeding as in the previous strategy one finds [86]

$$\xi_f^{(s)} \cdot \xi_{\bar{f}}^{(s)} = e^{-i2(2\beta_s+\gamma)} \tag{156}$$

with $\xi_f^{(s)}$ and $\xi_{\bar{f}}^{(s)}$ being the analogs of $\xi_f^{(d)}$ and $\xi_{\bar{f}}^{(d)}$, respectively. Now, all interferring amplitudes are of a similar size. With β_s extracted one day from the asymmetry in $B_s^0(\bar{B}_s^0) \to J/\psi\phi$, the angle γ can be determined.

$B^{\pm} \to D^0 K^{\pm}$, $B^{\pm} \to \bar{D}^0 K^{\pm}$ and γ. By replacing the spectator s-quark in the last strategy through the u-quark one arrives at decays of $B^{\pm}$ that can be used to extract γ. Also this strategy is unaffected by penguin contributions. Moreover, as particle-antiparticle mixing is absent here, γ can be measured directly without any need for phases in the mixing. Both these features make it plausible that this strategy, not involving to first approximation any loop diagrams, is particularly suited for the determination of γ without any new physics pollution.

By considering six decay rates $B^{\pm} \to D^0_{CP} K^{\pm}$, $B^+ \to D^0 K^+$, $\bar{D}^0 K^+$ and $B^- \to D^0 K^-$, $\bar{D}^0 K^-$ where $D^0_{CP} = (D^0 + \bar{D}^0)/\sqrt{2}$ is a CP eigenstate, and noting that

$$A(B^+ \to \bar{D}^0 K^+) = A(B^- \to D^0 K^-), \tag{157}$$

$$A(B^+ \to D^0 K^+) = A(B^- \to \bar{D}^0 K^-)e^{2i\gamma} \tag{158}$$

the well known triangle construction due to Gronau and Wyler [87] allows to determine γ. However, the method is not without problems. The detection of D^0_{CP}, that is necessary for this determination because $K^+ \bar{D}^0 \neq K^+ D^0$, is experimentally challenging. Moreover, the small branching ratios of the colour supressed channels in (158) and the absence of this suppression in the two remaining channels in (157) imply a rather squashed triangle thereby making the extraction of γ very difficult. Variants of this method that could be more promising are discussed in [88,89].

Other Clean Strategies for γ and β. The three strategies discussed above can be generalized to other decays. In particular [88,90]

- $2\beta + \gamma$ and γ can be measured in

$$B^0_d \to K_S D^0, \; K_S \bar{D}^0, \qquad B^0_d \to \pi^0 D^0, \; \pi^0 \bar{D}^0 \tag{159}$$

and the corresponding CP conjugated channels,
- $2\beta_s + \gamma$ and γ can be measured in

$$B^0_s \to \phi D^0, \; \phi \bar{D}^0, \qquad B^0_s \to K^0_S D^0, \; K_S \bar{D}^0 \tag{160}$$

and the corresponding CP conjugated channels,
- γ can be measured by generalizing the Gronau–Wyler construction to $B^{\pm} \to D^0 \pi^{\pm}, \bar{D}^0 \pi^{\pm}$ and to B_c decays [91]:

$$B^{\pm}_c \to D^0 D^{\pm}_s, \; \bar{D}^0 D^{\pm}_s, \qquad B^{\pm}_c \to D^0 D^{\pm}, \; \bar{D}^0 D^{\pm} . \tag{161}$$

In this context I can strongly recommend recent papers by Fleischer [90] that while discussing these decays go far beyond the methods presented here. It appears that the methods discussed in this subsection may give useful results at later stages of CP-B investigations, in particular at LHC-B and BTeV.

5.5   U–Spin Strategies

Preliminaries. Useful strategies for γ using the U-spin symmetry have been proposed in [92,93]. The first strategy involves the decays $B^0_{d,s} \to \psi K_S$ and $B^0_{d,s} \to D^+_{d,s} D^-_{d,s}$. The second strategy involves $B^0_s \to K^+ K^-$ and $B^0_d \to \pi^+ \pi^-$. They are unaffected by FSI and are only limited by U-spin breaking effects. They are promising for Run II at FNAL and in particular for LHC-B.

A method of determining γ, using $B^+ \to K^0 \pi^+$ and the U-spin related processes $B^0_d \to K^+ \pi^-$ and $B^0_s \to \pi^+ K^-$, was presented in [94]. A general discussion of U-spin symmetry in charmless B decays and more references to this topic can be found in [16,95]. I will only briefly discuss the method in [93].

$B^0_d \to \pi^+\pi^-$, $B^0_s \to K^+K^-$ and (γ, β). Replacing in $B^0_d \to \pi^+\pi^-$ the d quark by the s quark we obtain the decay $B^0_s \to K^+K^-$. The amplitude can be then written in analogy to (142) as follows

$$A(B^0_s \to K^+K^-) = V_{us}V^*_{ub}(A'_T + P'_u - P'_t) + V_{cs}V^*_{cb}(P'_c - P'_t). \qquad (162)$$

This formula differs from (142) only by $d \to s$ and the primes on the hadronic matrix elements that in principle are different in these two decays. As

$$V_{cs}V^*_{cb} \approx A\lambda^2, \qquad V_{us}V^*_{ub} \approx A\lambda^4 R_b e^{i\gamma}, \qquad (163)$$

the second term in (162) is even more important than the corresponding term in the case of $B^0_d \to \pi^+\pi^-$. Consequently $B^0_d \to K^+K^-$ taken alone does not offer a useful method for the determination of the CKM phases. On the other hand, with the help of the U-spin symmetry of strong interations, it allows roughly speaking to determine the penguin contributions in $B^0_d \to \pi^+\pi^-$ and consequently the extraction of β and γ.

Indeed, from the U-spin symmetry we have

$$\frac{P_{\pi\pi}}{T_{\pi\pi}} = \frac{P_c - P_t}{A_T + P_u - P_t} = \frac{P'_c - P'_t}{A'_T + P'_u - P'_t} = \frac{P_{KK}}{T_{KK}} \equiv de^{i\delta} \qquad (164)$$

where d is a real non-perturbative parameter and δ a strong phase. Measuring S_f and C_f for both decays and extracting β_s from $B^0_s \to J/\psi\phi$, we can determine four unknows: d, δ, β and γ subject mainly to U-spin breaking corrections. A recent analysis using these ideas can be found in [81].

5.6   Constraints for γ from $B \to \pi K$

Preliminaries. The recent developments involve also the extraction of the angle γ from the decays $B \to \pi K$. The modes $B^\pm \to \pi^\mp K^0$, $B^\pm \to \pi^0 K^\pm$, $B^0_d \to \pi^\mp K^\pm$ and $B^0_d \to \pi^0 K^0$ have been observed by the CLEO, BaBar and Belle collaborations and should allow us to obtain direct information on γ

when the errors on branching ratios and the CP asymmetries decrease. The latter are still consistent with zero. The progress on the accuracy of these measurements is slow but steady and they will certainly give an interesting insight into the flavour dynamics and QCD dynamics one day.

There has been a large theoretical activity in this field during the last six years. The main issues here are the final state interactions (FSI), SU(3) symmetry breaking effects and the importance of electroweak penguin contributions. Several interesting ideas have been put forward to extract the angle γ in spite of large hadronic uncertainties in $B \to \pi K$ decays [96–101].

Three strategies for bounding and determining γ have been proposed. The "mixed" strategy [96] uses $B_d^0 \to \pi^0 K^\pm$ and $B^\pm \to \pi^\pm K$. The "charged" strategy [101] involves $B^\pm \to \pi^0 K^\pm$, $\pi^\pm K$ and the "neutral" strategy [99] the modes $B_d^0 \to \pi^\mp K^\pm$, $\pi^0 K^0$. General parametrizations for the study of the FSI, SU(3) symmetry breaking effects and of the electroweak penguin contributions in these channels have been presented in [98–100]. Moreover, general parametrizations by means of Wick contractions [102,103] have been proposed. They can be used for all two-body B-decays. These parametrizations should turn out to be useful when the data improve.

Parallel to these efforts an important progress has been made by developing approaches for the calculation of the hadronic matrix elements of local operators in QCD beyond the standard factorization method. These are in particular the QCD factorization approach [104], the perturbative QCD approach [105] and the soft-collinear effective theory [106]. Moreover new methods to calculate exclusive hadronic matrix elements from QCD light-cone sum rules have been developed in [107]. While, in my opinion, an important progress in evaluating non-leptonic amplitudes has been made in these papers, the usefulness of this recent progress at the quantitative level has still to be demonstrated when the data improve.

A General Parametrization for $B \to \pi K$. In order to illustrate the complexity of the extraction of γ from these decays let me describe briefly the general parametrization for the mixed, charged and neutral strategies, developed in 1998 in collaboration with Robert Fleischer [99].

The isospin symmetry implies in each case one relation between the relevant amplitudes:

$$A(B^+ \to \pi^+ K^0) + A(B_d^0 \to \pi^- K^+) = -\left[T + P_{\mathrm{EW}}^{\mathrm{C}}\right] \tag{165}$$

$$A(B^+ \to \pi^+ K^0) + \sqrt{2} A(B^+ \to \pi^0 K^+) = -[(T + C) + P_{\mathrm{EW}}] = 3A_{3/2} \tag{166}$$

$$\sqrt{2} A(B_d^0 \to \pi^0 K^0) + A(B_d^0 \to \pi^- K^+) = -[(T + C) + P_{\mathrm{EW}}] = 3A_{3/2} \tag{167}$$

where T stands for tree, C for colour suppressed tree, P_{EW} for electroweak penguins and $P_{\mathrm{EW}}^{\mathrm{C}}$ for colour suppressed electroweak penguins. $A_{3/2}$ is an isospin amplitude. In particular we have

$$T + C = |T + C| e^{i\delta_{T+C}} e^{i\gamma}, \qquad P_{\mathrm{EW}} = -|P_{\mathrm{EW}}| e^{i\delta_{\mathrm{EW}}} \tag{168}$$

where the δ_i denote the strong interaction phases.

The QCD penguins, absent in (165)–(167), enter the analysis in the following manner:

$$P_{ch} \equiv A(B^+ \to \pi^+ K^0) = -(1 - \frac{\lambda^2}{2})\lambda^2 A \left[1 + \varrho_{ch} e^{i\theta_{ch}} e^{i\gamma}\right] |P_{tc}^{ch}| e^{i\delta_{tc}^{ch}} \quad (169)$$

$$P_n \equiv \sqrt{2}A(B_d^0 \to \pi^0 K^0) = -(1 - \frac{\lambda^2}{2})\lambda^2 A \left[1 + \varrho_n e^{i\theta_n} e^{i\gamma}\right] |P_{tc}^n| e^{i\delta_{tc}^n} \quad (170)$$

where the terms proportional to $\varrho_{ch,n}$ parametrize u-penguin and rescattering effects and the last factors stand for the difference of t and c penguins.

The relevant parameters in the three strategies in question are

$$r = \frac{|T|}{\sqrt{|P_{ch}|^2}}, \qquad q = \left|\frac{P_{EW}^C}{T}\right| e^{i\bar{\omega}}, \qquad \delta = \delta_T - \delta_{tc} \quad (171)$$

$$r_{ch} = \frac{|T+C|}{\sqrt{|P_{ch}|^2}}, \qquad q_{ch} = \left|\frac{P_{EW}}{T+C}\right| e^{i\omega}, \qquad \delta_{ch} = \delta_{T+C} - \delta_{tc}^{ch} \quad (172)$$

$$r_n = \frac{|T+C|}{\sqrt{|P_n|^2}}, \qquad q_n = q_{ch}, \qquad \delta_n = \delta_{T+C} - \delta_{tc}^n . \quad (173)$$

The virtue of this general parametrization is the universality of various formulae for quantities of interest, not shown here due to the lack of space. In order to study a given strategy, the relevant parameters listed above have to be inserted in these formulae.

The formulae given above are not sufficiently informative for a determination of γ. To proceed further one has to use $SU(3)$ flavour symmetry. This allows to fix r_{ch}, r_n and $q_{ch} = q_n$. On the other hand r and q are not determined by $SU(3)$ and their values have to be estimated by some dynamical assumptions like factorization. Consequently the mixed strategy has larger theoretical uncertainties than the other two strategies.

We have then, respectively, [101,108]

$$q_{ch} = q_n = 0.70 \left[\frac{0.37}{R_b}\right] = 0.70 \pm 0.08, \quad (174)$$

$$r_{ch} = \sqrt{2} \left|\frac{V_{us}}{V_{ud}}\right| \frac{F_K}{F_\pi} \sqrt{\frac{Br(B^\pm \to \pi^\pm \pi^0)}{Br(B^\pm \to \pi^\pm K^0)}} = 0.20 \pm 0.02 \quad (175)$$

with $|T+C|$ in r_{ch} extracted from $B^\pm \to \pi^\pm \pi^0$. The last number is my own estimate. Similarly one finds [99]

$$r_n = \left|\frac{V_{us}}{V_{ud}}\right| \frac{F_K}{F_\pi} \sqrt{\frac{Br(B^\pm \to \pi^\pm \pi^0)}{Br(B_d^0 \to \pi^0 K^0)}} \sqrt{\frac{\tau(B_d^0)}{\tau(B^\pm)}} = 0.19 \pm 0.02. \quad (176)$$

As demonstrated in a vast number of papers [96,98–101], these strategies imply interesting bounds on γ that not necessarily agree with the values extracted from the UT analysis of Sect. 3. In particular already in 1999 combining the neutral and charged strategies [99] we have found that the 1999 data on $B \to \pi K$ favour γ in the second quadrant, which is in conflict with the standard analysis of the unitarity triangle that implied $\gamma = (62\pm7)°$. Other arguments for $\cos\gamma < 0$ using $B \to PP$, PV and VV decays were given also in [109]. Recent analyses of $B \to \pi K$ by various authors find also that $\gamma > 90°$ is favoured by the $B \to \pi K$ data. Most recent reviews can be found in [3]. See also [110].

In view of sizable theoretical uncertainties in the analyses of $B \to \pi K$ and of still significant experimental errors in the corresponding branching ratios it is not yet clear whether the discrepancy in question is serious. For instance [111] sizable contributions of the so-called charming penguins to the $B \to \pi K$ amplitudes could shift γ extracted from these decays below 90° but at present these contributions cannot be calculated reliably. Similar role could be played by annihilation contributions [105] and large non-factorizable SU(3) breaking effects [99]. Also, new physics contributions in the electroweak penguin sector could shift γ to the first quadrant [99]. It should be however emphasized that the problem with the angle γ, if it persisted, would put into difficulties not only the SM but also the full class of MFV models in which the lower bound on $\Delta M_s/\Delta M_d$ implies $\gamma < 90°$. In any case it will be exciting to follow the developments in this field. The most recent analysis of this issue can be found in [112].

6 $K^+ \to \pi^+\nu\bar{\nu}$ and $K_L \to \pi^0\nu\bar{\nu}$

The rare decays $K^+ \to \pi^+\nu\bar{\nu}$ and $K_L \to \pi^0\nu\bar{\nu}$ are very promising probes of flavour physics within the SM and possible extensions, since they are governed by short distance interactions. They proceed through Z^0-penguin and box diagrams. As the required hadronic matrix elements can be extracted from the leading semileptonic decays and other long distance contributions turn out to be negligible [113], the relevant branching ratios can be computed to an exceptionally high degree of precision [114–116]. The main theoretical uncertainty in the CP conserving decay $K^+ \to \pi^+\nu\bar{\nu}$ originates in the value of $m_c(\mu_c)$. It has been reduced through NLO corrections down to $\pm7\%$ [114,115] at the level of the branching ratio. The dominantly CP-violating decay $K_L \to \pi^0\nu\bar{\nu}$ [117] is even cleaner as only the internal top contributions matter. The theoretical error for $Br(K_L \to \pi^0\nu\bar{\nu})$ amounts to $\pm2\%$ and is safely negligible.

6.1   Branching Ratios

The basic formulae for the branching ratios are given as follows

$$Br(K^+ \to \pi^+\nu\bar{\nu}) = \kappa_+ \cdot \left[(\mathrm{Im}F_t)^2 + (\mathrm{Re}F_c + \mathrm{Re}F_t)^2\right] , \qquad (177)$$

$$Br(K_L \to \pi^0 \nu\bar{\nu}) = \kappa_L \cdot (\mathrm{Im}F_t)^2 , \tag{178}$$

where

$$F_c = \frac{\lambda_c}{\lambda} P_0(X), \qquad F_t = \frac{\lambda_t}{\lambda^5} X(x_t). \tag{179}$$

Here $\lambda_i = V_{is}^* V_{id}$ and

$$\kappa_+ = 4.75 \cdot 10^{-11}, \qquad \kappa_L = 2.08 \cdot 10^{-10} \tag{180}$$

include isospin breaking corrections in relating $K^+ \to \pi^+ \nu\bar{\nu}$ and $K_L \to \pi^0 \nu\bar{\nu}$ to $K^+ \to \pi^0 e^+ \nu$, respectively [118]. Next

$$X(x_t) = 1.52 \cdot \left[\frac{\overline{m}_t(m_t)}{167 \ GeV} \right]^{1.15} \tag{181}$$

represents internal top contribution and $P_0(X) = 0.39 \pm 0.06$ results from the internal charm contribution [114]. The numerical values in (180) and for $P_0(X)$ differ from [115] due to a different value of $\lambda = 0.224$ used here.

Imposing all existing constraints on the CKM matrix one finds [119]

$$Br(K^+ \to \pi^+ \nu\bar{\nu}) = (7.7 \pm 1.1) \cdot 10^{-11}, \tag{182}$$

$$Br(K_L \to \pi^0 \nu\bar{\nu}) = (2.6 \pm 0.5) \cdot 10^{-11} \tag{183}$$

where the errors come dominantly from the uncertainties in the CKM parameters. Similar results are found in [120]. The first result should be compared with the measurements of AGS E787 collaboration at Brookhaven [121] that observing two events for this very rare decay finds

$$Br(K^+ \to \pi^+ \nu\bar{\nu}) = (15.7^{+17.5}_{-8.2}) \cdot 10^{-11} . \tag{184}$$

This is a factor of 2 above the SM expectation. Even if the errors are substantial and the result is compatible with the SM, the branching ratio (184) implies already a non-trivial lower bound on $|V_{td}|$ [121,122].

The present upper bound on $Br(K_L \to \pi^0 \nu\bar{\nu})$ from the KTeV experiment at Fermilab [123] reads

$$Br(K_L \to \pi^0 \nu\bar{\nu}) < 5.9 \cdot 10^{-7} . \tag{185}$$

This is about four orders of magnitude above the SM expectation (183). Moreover this bound is substantially weaker than the *model independent* bound [124] from isospin symmetry:

$$Br(K_L \to \pi^0 \nu\bar{\nu}) < 4.4 \cdot Br(K^+ \to \pi^+ \nu\bar{\nu}) \tag{186}$$

which through (184) gives

$$Br(K_L \to \pi^0 \nu\bar{\nu}) < 1.6 \cdot 10^{-9} \ (90\% C.L.) \tag{187}$$

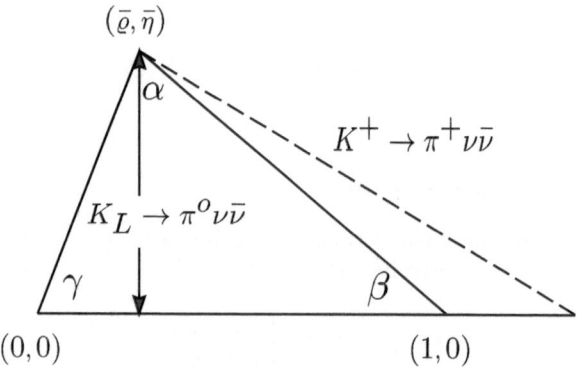

Fig. 9. Unitarity triangle from $K \to \pi\nu\bar{\nu}$.

6.2   Unitarity Triangle and $\sin 2\beta$ from $K \to \pi\nu\bar{\nu}$

The measurement of $Br(K^+ \to \pi^+\nu\bar{\nu})$ and $Br(K_{\mathrm{L}} \to \pi^0\nu\bar{\nu})$ can determine the unitarity triangle completely, (see Fig. 9) [82]. The explicit formulae can be found in [21,6,82]. The most interesting in this context are very clean determinations of $\sin 2\beta$ and $\mathrm{Im}\lambda_t$ that are free not only from hadronic uncertainties but also parametric uncertainties like $|V_{cb}|$ and m_c. The determination of $|V_{td}|$ is also theoretically clean but its precision depends on the accuracy with which $|V_{cb}|$ and m_c are known. Also the scale uncertainties in $|V_{td}|$ amount to 4% at the NLO [114]. They should be significantly reduced through a calculation of NNLO corrections to the charm contribution that is in progress and should be available in 2004.

Assuming that the branching ratios will be known to within $\pm 10\%$ we expect the following accuracy in this decade

$$\sigma(\sin 2\beta) = \pm 0.04, \quad \sigma(\mathrm{Im}\lambda_t) = \pm 5\%, \quad \sigma(|V_{td}|) = \pm 7\% . \tag{188}$$

The comparison with the corresponding determinations in B decays will offer a very good test of flavour dynamics and CP violation in the SM and a powerful tool to probe the physics beyond it.

6.3   Concluding Remarks

As the theorists were able to calculate the branching ratios for these decays rather precisely, the future of this field is in the hands of experimentalists and depends on the financial support that is badly needed. The experimental outlook for these decays has been reviewed in [125,126]. The future of $K^+ \to \pi^+\nu\bar{\nu}$ depends on the AGS E949 and the CKM experiment at Fermilab. In the case of $K_{\mathrm{L}} \to \pi^0\nu\bar{\nu}$ these are the KEK E391a experiment, KOPIO at

Brookhaven (BNL E926) and an experiment at the 50 GeV JHF in Japan that should be able to collect 1000 events at the end of this decade. Both KOPIO and JHF should provide very important measurements of this gold-plated decay. For a recent theoretical review see [119].

7   Minimal Flavour Violation Models

7.1   Preliminaries

We have defined this class of models in Sect. 1. Here I would like just to list four interesting properties of these models that are independent of particular parameters present in these models. Other relations can be found in [127]. These are:

- There exists a universal unitarity triangle (UUT) [24] common to all these models and the SM that can be constructed by using measurable quantities that depend on the CKM parameters but are not polluted by the new parameters present in the extensions of the SM. The UUT can be constructed, for instance, by using $\sin 2\beta$ from $a_{\psi K_S}$ and the ratio $\Delta M_s / \Delta M_d$. The relevant formulae can be found in Sect. 3 and in [24,128], where also other quantities suitable for the determination of the UUT are discussed.

-

$$(\sin 2\beta)_{J/\psi K_S} = (\sin 2\beta)_{\phi K_S} = (\sin 2\beta)_{\pi\nu\bar\nu} \tag{189}$$

- For given $a_{\psi K_S}$ and $Br(K^+ \to \pi^+ \nu\bar\nu)$ only two values of $Br(K_L \to \pi^0 \nu\bar\nu)$ are possible in the full class of MFV models, independently of any new parameters present in these models [128]. Consequently, measuring $Br(K_L \to \pi^0 \nu\bar\nu)$ will either select one of these two possible values or rule out all MFV models. The present experimental bound on $Br(K^+ \to \pi^+ \nu\bar\nu)$ and $\sin 2\beta \le 0.80$ imply an absolute upper bound $Br(K_L \to \pi^0 \nu\bar\nu) < 4.9 \cdot 10^{-10}$ (90% C.L.) [128] in the MFV models that is stronger than the bound in (187).

- There exists a correlation between $Br(B_{d,s} \to \mu\bar\mu)$ and $\Delta M_{d,s}$ [129]:

$$\frac{Br(B_s \to \mu\bar\mu)}{Br(B_d \to \mu\bar\mu)} = \frac{\hat{B}_d}{\hat{B}_s} \frac{\tau(B_s)}{\tau(B_d)} \frac{\Delta M_s}{\Delta M_d} \tag{190}$$

that is practically free of theoretical uncertainties as $\hat{B}_s / \hat{B}_d = 1$ up to small $SU(3)$ breaking corrections. Similar correlations between $Br(B_{d,s} \to \mu\bar\mu)$ and $\Delta M_{d,s}$, respectively, allow rather precise predictions for $Br(B_{d,s} \to \mu\bar\mu)$ within the MFV models once $\Delta M_{d,s}$ are known [129].

7.2   Universal Unitarity Triangle

The presently available quantities that do not depend on the new physics parameters within the MFV models and therefore can be used to determine the UUT are R_t from $\Delta M_d / \Delta M_s$ by means of (104), R_b from $|V_{ub}/V_{cb}|$ by means of (12) and $\sin 2\beta$ extracted from the CP asymmetry in $B_d^0 \to \psi K_S$. Using only these three quantities, we show in Fig. 6 the allowed universal region for $(\bar{\varrho}, \bar{\eta})$ (the larger ellipse) in the MFV models as obtained recently in an update of [44]. The results for various quantities of interest related to this UUT are collected in Table 1. Similar analysis has been done in [26].

It should be stressed that any MFV model that is inconsistent with the broader allowed region in Fig. 6 and the UUT column in Table 1 is ruled out. We observe that there is little room for MFV models that in their predictions for UT differ significantly from the SM. It is also clear that to distinguish the SM from the MFV models on the basis of the analysis of the UT of Sect. 3, will require considerable reduction of theoretical uncertainties.

7.3   Models with Universal Extra Dimensions

In view of the difficulty in distinguishing various MFV models on the basis of the standard analysis of UT from each other, it is essential to study other FCNC processes as rare B and K decays and radiative B decays like $B \to X_s \gamma$ and $B \to X_s \mu^+ \mu^-$. In the case of MSSM at low $\tan \beta$ such an analyses can be found in [50,130]. Recently a very extensive analysis of all relevant FCNC processes in a SM with one universal extra dimension [131] has been presented in [132,133]. In this model all standard model fields can propagate in the fifth dimension and the FCNC processes are affected by the exchange of the Kaluza-Klein particles in loop diagrams. The most interesting results of [132,133] are the enhancements of $Br(K^+ \to \pi^+ \nu \bar{\nu})$ and $Br(B \to X_s \mu^+ \mu^-)$, strong suppressions of $Br(B \to X_s \gamma)$ and $Br(B \to X_s$ gluon$)$ and a significant downward shift of the zero $\hat{s}_0$ in the forward-backward asymmetry in $Br(B \to X_s \mu^+ \mu^-)$.

8   Outlook

Let me finish these lectures with my personal expectations for the coming years with regard to the CKM matrix and FCNC processes.

8.1   Phase 1 (2003-2007)

In this phase the determination of the CKM matrix will be governed by

$$V_{us}, \qquad |V_{cb}|, \qquad a_{\psi K_S}, \qquad \Delta M_d / \Delta M_s. \tag{191}$$

These four quantities are sufficient to determine the full CKM matrix and suggest a new set of fundamental variables [44]

$$V_{us}, \qquad |V_{cb}|, \qquad \beta, \qquad R_t. \qquad (192)$$

The precision of this determination will depend on the accuracy with which $a_{\psi K_S}$ and $\Delta M_d / \Delta M_s$ will be measured and the non-perturbative ratio ξ calculated by lattice and QCD sum rules methods.

An important role will also be played by

$$\varepsilon_K, \qquad |V_{ub}/V_{cb}| \qquad (193)$$

but this will depend on the reduction of the hadronic uncertainties in $\hat{B}_K$ and in the determination of $|V_{ub}|$ [3]. Note that ε_K and $|V_{ub}/V_{cb}|$ combined with (191) can tell us whether the CP violation in the K-system is consistent with the one observed in the B-system.

Very important is the clarification of the possible discrepancy between the measurements of the angle β by means of $a_{J/\psi K_S}$ and $a_{\phi K_S}$ that if confirmed would imply new sources of CP violation. Similarly the status of CP violation in $B_d^0 \to \pi^+\pi^-$ and in $B \to \pi K$ decays should be clarified in this phase but this will depend on the theoretical progress in non-perturbative methods. We should also be able to get some information about γ not only from $B \to \pi K$ at B factories but also by means of U-spin strategies in conjuction with the data from Run II at Tevatron.

During this phase we should also have new data on $Br(K^+ \to \pi^+\nu\bar{\nu})$ from AGS E949 and first data from the CKM experiment at Fermilab. The comparison of these data with the implications of $\Delta M_d / \Delta M_s$ should be very interesting [115,122,132,81]. It would be particularly exciting if the central value did not decrease below the one in (184), that is roughly by a factor of two higher than the SM value. In any case these data should have a considerable impact on $|V_{td}|$ and the unitarity triangle.

We will also have new data on $B_s \to \mu^+\mu^-$, $B \to X_s\nu\bar{\nu}$, $B \to X_s\gamma$, $B \to X_s\mu\bar{\mu}$ as well as on related exclusive channels. All these decays are governed by the CKM element $|V_{ts}|$ that is already well determined by the unitarity of the CKM matrix $|V_{ts}| \approx |V_{cb}|$. Consequently I do not expect that these decays will play an important role in the CKM fits. On the other hand being sensitive to new physics contributions they could give the first signals of new physics. The fact that $|V_{ts}|$ is already reasonably well known will be helpful in this context.

8.2   Phase 2 (2007-2009)

With the B-factories and Tevatron entering their mature stage and LHCB, BTeV, Atlas and CMS beginning hopefully their operation, the quantities in (191) should offer a very good determination of the CKM matrix. I expect that other decays listed in Phase 1 will become more useful in view of improved data and new theoretical ideas. The most important new developments to be expected in this phase will be clean measurements of the angle γ at LHCB and BTeV in decays $B_s \to D_s^+ K^-$ and $\bar{B}_s \to D_s^- K^+$ and an improved measurement of $Br(K^+ \to \pi^+ \nu \bar{\nu})$ by the CKM collaboration at Fermilab. Also other strategies discussed in Sect. 5.4 and possible measurements of rare B-decays sensitive to both $|V_{ts}|$ and $|V_{td}|$ should play an important role. This phase should provide (in case the phase 1 did not do it) definite answer whether MFV is sufficient to describe the data or whether new flavour violating interactions are required.

At the end of this phase we should also have much more improved knowledge about $K_L \to \pi^0 \nu \bar{\nu}$ from KOPIO at Brookhaven and JHF in Japan.

However, the most interesting scenario would be the discovery of supersymmetry at LHC which could considerably reduce the uncertainty in the supersymmetric parameters necessary for the study of FCNC processes.

8.3   Phase 3 (2009-2013)

Here precise measurements of $Br(K_L \to \pi^0 \nu \bar{\nu})$ from KOPIO and JHF will be among the highlights. In addition the branching ratios for most of the decays studied in phases 1 and 2 will be known with much higher precision. This will allow not only a precision test of SM but also to identify the patterns of new physics contributions that I personnally expect should show up at this level of accuracy. The combination of these studies with the results from LHC that should signal some direct signs of new physics should allow a convincing identification of this new physics.

No doubt the next ten years should be very exciting but the real progress will require extreme joined efforts by theorists and experimentalists.

Acknowledgements

I would like to thank the organizers for inviting me to such a wonderful winter school and most enjoyable atmosphere. I would also like to thank Robert Fleischer, Matthias Jamin, Stefan Recksiegel and Achille Stocchi for discussions. The work presented here has been supported in part by the German Bundesministerium für Bildung und Forschung under the contract 05HT1WOA3 and the DFG Project Bu. 706/1-2.

References

1. N. Cabibbo, Phys. Rev. Lett. **10**, 531 (1963)
2. M. Kobayashi and K. Maskawa, Prog. Theor. Phys. **49**, 652 (1973)
3. M. Battaglia, A.J. Buras, P. Gambino, A. Stocchi *et al.*, hep-ph/0304132

4. A.J. Buras, hep-ph/9806471, in *Probing the Standard Model of Particle Interactions*, eds. R. Gupta, A. Morel, E. de Rafael and F. David (Elsevier Science B.V., Amsterdam, 1998) pp. 281–539

5. A.J. Buras, hep-ph/9905437, in *Electroweak Physics*, eds. A. Astbury et al, (World Scientific 2000) pp. 1–83.

6. A.J. Buras, hep-ph/0101336, in proceedings of the International School of Subnuclear Physics, ed. A .Zichichi, (World Scientific 2001) pp. 200–337 2000.

7. A.J. Buras, AIP Conf. Proc. **623**, 3 (2002)

8. G. Buchalla, A.J. Buras and M. Lautenbacher, Rev. Mod. Phys **68**, 1125 (1996)

9. A.J. Buras and R. Fleischer, Adv. Ser. Direct. High. Energy Phys. **15**, 65 (1998)

10. Heavy Flavours II, eds. A.J. Buras and M. Lindner, World Scientific, (1998)

11. G. Branco, L. Lavoura and J. Silva, (1999), CP Violation, Oxford Science Publications, Clarendon Press, Oxford

12. I.I. Bigi and A.I. Sanda, (2000), CP Violation, Cambridge Monographs on Particle Physics, Nuclear Physics and Cosmology, Cambridge University Press, Cambridge.

13. The BaBar Physics Book, eds. P. Harrison and H. Quinn, (1998), SLAC report 504.

14. B Decays at the LHC, eds. P. Ball, R. Fleischer, G.F. Tartarelli, P. Vikas and G. Wilkinson, hep-ph/0003238.

15. B Physics at the Tevatron, Run II and Beyond, K. Anikeev et al., hep-ph/0201071.

16. R. Fleischer, Phys. Rep. **370**, 537 (2002) Y. Nir, hep-ph/9911321, hep-ph/0109090 G. Buchalla, hep-ph/020292, hep-ph/0302145

17. H. Fritzsch and Z.Z. Xing, Prog. Part. Nucl. Phys. **45**, 1 (2000)

18. L.L. Chau and W.-Y. Keung, Phys. Rev. Lett. **53**, 1802 (1984)

19. K. Hagiwara et al., "Review of Particle Physics", Phys. Rev. **D 66**, 010001 (2002)

20. L. Wolfenstein, Phys. Rev. Lett. **51**, 1945 (1983)

21. A.J. Buras, M.E. Lautenbacher and G. Ostermaier, Phys. Rev. **D 50**, 3433 (1994)

22. M. Schmidtler and K.R. Schubert, Z. Phys. **C 53**, 347 (1992)

23. R. Aleksan, B. Kayser and D. London, Phys. Rev. Lett. **73**, 18 (1994)

24. A.J. Buras, P. Gambino, M. Gorbahn, S. Jäger and L. Silvestrini, Phys. Lett. **D500**, 161 (2001)

25. C. Bobeth, T. Ewerth, F. Krüger and J. Urban, Phys. Rev. **D66**, 074021 (2002)

26. G. D'Ambrosio, et al., Nucl. Phys. **B645**, 155 (2002)

27. T. Inami and C.S. Lim, Progr. Theor. Phys. **65**, 297 (1981)

28. A.J. Buras, hep-ph/0109197, in proceedings of *Kaon 2001*, eds. F. Costantini, G. Isodori, M. Sozzi, pp. 15–43

29. S.L. Glashow, J. Iliopoulos and L. Maiani Phys. Rev. **D2**, 1285 (1970)

30. J. Bijnens, J.-M. Gérard and G. Klein, Phys. Lett. **B257**, 191 (1991)

31. S. Herrlich and U. Nierste, Nucl. Phys. **B419**, 292 (1994)

32. L.L. Chau, Physics Reports, **95**, 1 (1983)

33. J. Gasser and U.G. Meißner, Phys. Lett. **B258**, 219 (1991)

34. A.J. Buras, M. Jamin, and P.H. Weisz, Nucl. Phys. **B347**, 491 (1990)

35. S. Herrlich and U. Nierste, Phys. Rev. **D52**, 6505 (1995) Nucl. Phys. **B476**, 27 (1996)

36. M. Jamin and U. Nierste, recent update.

37. J. Urban, F. Krauss, U. Jentschura and G. Soff, Nucl. Phys. **B523**, 40 (1998)

38. The numerical value on the r.h.s of (100) being proportional to λ^{-10} is very sensitive to λ but as $A \propto \lambda^{-2}$ and $P_c(\varepsilon) \propto \lambda^{-4}$ the constraint (100) depends only quadratically on λ.

39. B. Aubert et al., Phys. Rev. Lett. **89**, 201802 (2002)

40. K. Abe et al., Phys. Rev. **D66**, 071102 (2002)

41. Y. Nir, Nucl. Phys. Proc. Suppl. **117**, 111 (2003)

42. M. Ciuchini et al., JHEP **0107**, 013 (2001)

43. A. Höcker, et al., Eur. Phys. J. **C21**, 225 (2001)

44. A.J. Buras, F. Parodi and A. Stocchi, JHEP **0301**, 029 (2003)

45. A. Lai et al., Eur. Phys. J. **C22**, 231 (2001) J.R. Batley et al., Phys. Lett. **B544**, 97 (2002)

46. A. Alavi-Harati et al., Phys. Rev. Lett. **83**, 22 (1999) Phys. Rev. **D67**, 012005 (2003)

47. H. Burkhardt et al., Phys. Lett. **B206**, 169 (1988) G.D. Barr et al., Phys. Lett. **B317**, 233 (1993)

48. L.K. Gibbons et al., Phys. Rev. Lett. **70**, 1203 (1993)

49. S. Bosch et al., Nucl. Phys. **B 565**, 3 (2000)

50. A.J. Buras et al., Nucl. Phys. **B592**, 55 (2001)

51. M. Ciuchini and G. Martinelli, Nucl. Phys. Proc. Suppl. **99B**, 27 (2001)

52. T. Hambye et al., Nucl. Phys. **B564**, 391 (2000)

53. S. Bertolini, J.O. Eeg and M. Fabbrichesi, Phys. Rev. **D63**, 056009 (2001)

54. E. Pallante and A. Pich, Phys. Rev. Lett. **84**, 2568 (2000) Nucl. Phys. **B592**, 294 (2001) E. Pallante, A. Pich and I. Scimemi, Nucl. Phys. **B617**, 441 (2001)

55. J. Bijnens and J. Prades, JHEP **06**, 035 (2000), Nucl. Phys. Proc. Suppl. **96**, 354 (2001) J. Bijnens, E. Gamiz and J. Prades, JHEP **10**, 009 (2001)

56. M. Knecht, S. Peris and E. de Rafael, Phys. Lett. **B508**, 117 (2001) T. Hambye, S. Peris and E. de Rafael, hep-ph/0305104.

57. S. Bertolini, M. Fabbrichesi and J.O. Eeg, Rev. Mod. Phys. **72**, 65 (2000)

58. S. Bertolini, hep-ph/0206095.

59. A.J. Buras, M. Jamin, M.E. Lautenbacher and P.H. Weisz, Nucl. Phys. **B370**, 69 (1992) Nucl. Phys. **B400**, 37 (1993) A.J. Buras, M. Jamin and M.E. Lautenbacher, Nucl. Phys. **B400**, 75 (1993)

60. A.J. Buras, M. Jamin and M.E. Lautenbacher, Nucl. Phys. **B408**, 209 (1993)

61. M. Ciuchini, E. Franco, G. Martinelli and Reina, Phys. Lett. **B301**, 263 (1993) Nucl. Phys. **B415**, 403 (1994)

62. A.J. Buras and M. Jamin, hep-ph/0306217

63. W.A. Bardeen, A.J. Buras and J.-M. Gérard, Phys. Lett. **B180**, 133 (1986) Nucl. Phys. **B293**, 787 (1987) Phys. Lett. **B192**, 138 (1987) Phys. Lett. **B211**, 343 (1988)

64. G. Buchalla, A.J. Buras and M.K. Harlander, Nucl. Phys. **B 349**, 1 (1991)

65. V. Cirigliano, A. Pich, G. Ecker, and H. Neufeld, hep-ph/0307030

66. Y.-Y. Keum, U. Nierste and A.I. Sanda, Phys. Lett. **B457**, 157 (1999)

67. RBC Collaboration, T. Blum et al., hep-lat/0110075; CP-PACS Collaboration, S. Aoki et al., Nucl. Phys. Proc. Suppl. **106B**, 332 (2002)

68. SPQCDR Collaboration, D. Becirevic et al., hep-lat/0209136

69. V. Cirigliano et al., Phys. Lett. **B555**, 71 (2003)
70. S. Narison, Nucl. Phys. **B593**, 3 (2001)
71. A.J. Buras and J.-M. Gérard, Phys. Lett. **B517**, 129 (2001)
72. J.F Donoghue, in proceedings of *Kaon 2001*, eds. F. Costantini, G. Isodori, M. Sozzi, pp 93–106
73. J. Hein, C. Davies, G.P. Lepage, Q. Mason and H. Trottier, hep-lat/0209077
74. R. Gupta, talk presented at *CKM matrix and Unitarity Triangle*, Durham, April 2003
75. Y. Grossman, G. Isidori and M.P. Worah, Phys. Rev. **D58**, 057504 (1998)
76. B. Aubert et al., hep-ex/020770
77. K. Abe et al., hep-ex/0207098
78. R. Fleischer and T. Mannel, Phys. Lett. **B511**, 240 (2001) G. Hiller, Phys. Rev. **D66**, 071502 (2002) A. Datta, Phys. Rev. **D66**, 071702 (2002) M. Ciuchini and L. Silvestrini, Phys. Rev. Lett. **89**, 231802 (2002) M. Raidal, Phys. Rev. Lett. **89**, 231803 (2002) Y. Grossman, Z. Ligeti, Y. Nir and H. Quinn, hep-ph/0303171 S. Khalil and E. Kou, hep-ph/0307024
79. M. Gronau and D. London, Phys. Rev. Lett. **65**, 3381 (1990)
80. R. Fleischer and J. Matias, Phys. Rev. **D66**, 054009 (2002) M. Gronau and J.L. Rosner, Phys. Rev. **D65**, 093012 (2002) D. Atwood and A. Soni, hep-ph/0206045 M. Ciuchini et al., Nucl. Phys. **B 501**, 271 (1997)
81. R. Fleischer, G. Isidori and J. Matias, JHEP **0305**, 053 (2003)
82. G. Buchalla and A.J. Buras, Phys. Lett. **B333**, 221 (1994) Phys. Rev. **D54**, 6782 (1996)
83. M. Beneke, G. Buchalla, M. Neubert and C.T. Sachrajda, Nucl. Phys. **B 606**, 245 (2001) M. Gronau and J.L. Rosner, Phys. Rev. **D65**, 013004 (2002) Z. Luo and J.L. Rosner, Phys. Rev. **D54**, 054027 (2002)
84. R.G. Sachs, EFI-85-22 (unpublished); I. Dunietz and R.G. Sachs, Phys. Rev. **D37**, 3186 (1988) [E: Phys. Rev. **D39**, 3515 (1988)] I. Dunietz, Phys. Lett. **427**, 179 (1998) M. Diehl and G. Hiller, Phys. Lett. **517**, 125 (2001)
85. J.P. Silva et al., Phys. Rev. **D67**, 036004 (2003)
86. R. Aleksan, I. Dunietz and B. Kayser, Z. Phys. **C54**, 653 (1992) R. Fleischer and I. Dunietz, Phys. Lett. **B387**, 361 (1996) A.F. Falk and A.A. Petrov, Phys. Rev. Lett. **85**, 252 (2000) D. London, N. Sinha and R. Sinha, Phys. Rev. Lett. **85**, 1807 (2000)
87. M. Gronau and D. Wyler, Phys. Lett. **B265**, 172 (1991)
88. M. Gronau and D. London, Phys. Lett. **B253**, 483 (1991) I. Dunietz, Phys. Lett. **B270**, 75 (1991)
89. D. Atwood, I. Dunietz and A. Soni, Phys. Rev. Lett. **B78**, 3257 (1997)
90. R. Fleischer, Nucl. Phys. **B659**, 321 (2003), Phys. Lett. **B562**, 234 (2003), hep-ph/0304027.
91. R. Fleischer and D. Wyler, Phys. Rev. **D62**, 057503 (2000)
92. R. Fleischer, Eur. Phys. J. **C10**, 299 (1999) Phys. Rev. **D60**, 073008 (1999)
93. R. Fleischer, Phys. Lett. **B459**, 306 (1999)
94. M. Gronau and J.L. Rosner, Phys. Lett. **B482**, 71 (2000) C.W. Chiang and L. Wolfenstein, Phys. Lett. **B493**, 73 (2000)
95. M. Gronau, Phys. Lett. **B492**, 297 (2000)
96. R. Fleischer, Phys. Lett. **B365**, 399 (1996) R. Fleischer and T. Mannel, Phys. Rev., **D57** 2752 (1998)
97. M. Gronau and J.L. Rosner, Phys. Rev. **D57**, 6843 (1998)

98. R. Fleischer, Eur. Phys. J. **C6**, 451 (1999)

99. A.J. Buras and R. Fleischer, Eur. Phys. J. **C11**, 93 (1999) Eur. Phys. J. **C16**, 97 (2000)

100. M. Neubert, JHEP **9902**, 014 (1998)

101. M. Neubert and J.L. Rosner, Phys. Lett. **B441**, 403 (1998) Phys. Rev. Lett. **81**, 5076 (1998)

102. M. Ciuchini et al., Nucl. Phys. **B501**, 271 (1997)

103. A.J. Buras and L. Silvestrini, Nucl. Phys. **B569**, 3 (2000)

104. M. Beneke, G. Buchalla, M. Neubert and C.T. Sachrajda, Phys. Rev. Lett. **83**, 1914 (1999) Nucl. Phys. **B 591**, 313 (2000) hep-ph/0007256; Nucl. Phys., **B606** 245 (2001)

105. H. -Y. Cheng, H.-n. Li and K.-C. Yang Phys. Rev. **D60**, 094005 (1999) Y.-Y. Keum, H.-n. Li and A.I. Sanda, Phys. Lett. **B504**, 6 (2001) Phys. Rev. **D63**, 054008 (2001)

106. Ch.W. Bauer et al., Phys. Rev. **D65**, 054022 (2002) Phys. Rev. **D66**, 014017 (2002) Phys. Rev. **D66**, 054005 (2002) I.W. Stewart, hep-ph/0208034 hep-ph/0109045 M. Beneke et al., Nucl. Phys. **B 643**, 431 (2002) Phys. Lett. **B553**, 267 (2003)

107. A. Khodjamirian, Nucl. Phys. **B 605**, 558 (2001) hep-ph/0108205 R. Rückl, S. Weinzierl and O. Yakovlev, hep-ph/0105161, hep-ph/0007344. A. Khodjamirian, T. Mannel and P. Urban, Phys. Rev. **D67**, 054027 (2003) A. Khodjamirian, T. Mannel and B. Melic, hep-ph/0304179

108. M. Gronau, J.l. Rosner and D. London, Phys. Rev. Lett. **73**, 21 (1994)

109. D. Cronin-Hennessy et al., (CLEO), Phys. Rev. Lett. **85**, 515 and 525 (2000) X.-G. He, W.-S. Hou and K-Ch. Yang, hep-ph/9902256 W.-S. Hou and K.-Ch. Yang, Phys. Rev. **D61**, 073014 (2000) W.-S. Hou, J.G. Smith and F. Würthwein, hep-ex/9910014.

110. M. Neubert, hep-ph/0207327.

111. M. Ciuchini et al., Phys. Lett. **515**, 33 (2001)

112. A.J. Buras, R. Fleischer, S. Recksiegel and F. Schwab, hep-ph/0309012.

113. D. Rein and L.M. Sehgal, Phys. Rev. **D39**, 3325 (1989) J.S. Hagelin and L.S. Littenberg, Prog. Part. Nucl. Phys. **23**, 1 (1989) M. Lu and M.B. Wise, Phys. Lett. **B324**, 461 (1994) S. Fajfer, [hep-ph/9602322]; C.Q. Geng, I.J. Hsu and Y.C. Lin, Phys. Rev. **D54**, 877 (1996) G. Buchalla and G. Isidori, Phys. Lett. **B440**, 170 (1998) A.F. Falk, A. Lewandowski and A.A. Petrov, Phys. Lett. **B505**, 107 (2001)

114. G. Buchalla and A.J. Buras, Nucl. Phys. **B 400**, 225 (1993) Nucl. Phys. **B 412**, 106 (1994)

115. G. Buchalla and A.J. Buras, Nucl. Phys. **B 548**, 309 (1999)

116. M. Misiak and J. Urban, Phys. Lett. **B541**, 161 (1999)

117. L. Littenberg, Phys. Rev. **D39**, 3322 (1989)

118. W. Marciano and Z. Parsa, Phys. Rev. **D53**, R1 (1996).

119. G. Isidori, hep-ph/0307014.

120. S. Kettell, L. Landsberg and H. Nguyen, hep-ph/0212321.

121. S. Adler et al., Phys. Rev. Lett. **79**, 2204 (1997) Phys. Rev. Lett. **84**, 3768 (2000) Phys. Rev. Lett. **88**, 041803 (2002)

122. G. D'Ambrosio and G. Isidori, Phys. Lett. **B530**, 108 (2002) hep-ph/0112135.

123. A. Alavi-Harati et al., Phys. Rev. **D61**, 072006 (2000)

124. Y. Grossman and Y. Nir, Phys. Lett. **B398**, 163 (1997)

125. L. Littenberg, hep-ex/0010048; T.K. Komatsubara, hep-ex/0112016.
126. A. Belyaev et al, hep-ph/0107046.
127. A.J. Buras and R. Buras, Phys. Lett. **B501**, 223 (2001) S. Bergmann and G. Perez, Phys. Rev. **D64**, 115009 (2001) JHEP **0008**, 034 (2000) S. Laplace, Z. Ligeti, Y. Nir and G. Perez, Phys. Rev. **D65**, 094040 (2002)
128. A.J. Buras and R. Fleischer, Phys. Rev. **D64**, 115010 (2001)
129. A.J. Buras, hep-ph/303060.
130. A. Ali et al., Phys. Rev. **D 66**, 034002 (2002)
131. T. Appelquist et al., Phys. Rev. **D 64**, 035002 (2001)
132. A.J. Buras, M. Spranger and A. Weiler, Nucl. Phys. **B 660**, 225 (2003)
133. A.J. Buras, A. Poschenrieder, M. Spranger and A. Weiler, hep-ph/0306158

Heavy-Quark Physics

Matthias Neubert

Institute for High-Energy Phenomenology, Newman Laboratory
for Elementary-Particle Physics, Cornell University, Ithaca, NY 14853, U.S.A.,
neubert@lepp.cornell.edu

Abstract. These lectures provide a pedagogical introduction to the theory of non-leptonic heavy-meson decays recently proposed by Beneke, Buchalla, Sachrajda and myself, which provides a rigorous basis for factorization for a large class of non-leptonic two-body B-meson decays in the heavy-quark limit. Some phenomenological applications of this approach are briefly discussed.

1 Introduction

Non-leptonic two-body decays of B mesons, although simple as far as the underlying weak decay of the b quark is concerned, are complicated on account of strong-interaction effects. If these effects could be computed, this would enhance tremendously our ability to uncover the origin of CP violation in weak interactions from data on a variety of such decays being collected at the B factories. In these lecture, I review recent progress toward a systematic analysis of weak heavy-meson decays into two energetic mesons based on the factorization properties of decay amplitudes in QCD [1–3]. The first part of my lectures follows closely the detailed account of this approach given in [2].

As in the classic analysis of semi-leptonic $B \to D$ transitions [4,5], our arguments make extensive use of the fact that the b quark is heavy compared to the intrinsic scale of strong interactions. This allows us to deduce that non-leptonic decay amplitudes in the heavy-quark limit have a simple structure. The arguments to reach this conclusion, however, are quite different from those used for semi-leptonic decays, since for non-leptonic decays a large momentum is transferred to at least one of the final-state mesons. The results of our work justify naive factorization of four fermion operators for many, but not all, non-leptonic decays and imply that corrections termed "non-factorizable", which up to now have been thought to be intractable, can be calculated rigorously if the mass of the decaying quark is large enough. This leads to a large number of predictions for branching fractions and CP asymmetries of two-body B decays, including many charmless decay modes, which are currently being studied at the B factories.

Weak decays of heavy mesons involve three fundamental scales, the weak-interaction scale M_W, the b-quark mass m_b, and the QCD scale $\Lambda_{\rm QCD}$, which are strongly ordered: $M_W \gg m_b \gg \Lambda_{\rm QCD}$. The underlying weak decay being computable, all theoretical work concerns strong-interaction correc-

M. Neubert, Heavy-Quark Physics, Lect. Notes Phys. **629**, 137–168 (2004)

tions. QCD effects involving virtualities above the scale m_b are well understood. They renormalize the coefficients of local operators O_i in the effective weak Hamiltonian [6], so that the amplitude for the decay $B \to M_1 M_2$ is given by

$$\mathcal{A}(B \to M_1 M_2) = \frac{G_F}{\sqrt{2}} \sum_i \lambda_i \, C_i(\mu) \, \langle M_1 M_2 | O_i(\mu) | B \rangle \,, \tag{1}$$

where each term in the sum is the product of a Cabibbo–Kobayashi–Maskawa (CKM) factor λ_i, a coefficient function $C_i(\mu)$, which incorporates strong-interaction effects above the scale $\mu \sim m_b$, and a matrix element of an operator O_i. The difficult theoretical problem is to compute these matrix elements or, at least, to reduce them to simpler non-perturbative objects.

A variety of treatments of this problem exist, which rely on assumptions of some sort. Here we identify two somewhat contrary lines of approach. The first one, which we shall call "naive factorization", replaces the matrix element of a four-fermion operator in a heavy-quark decay by the product of the matrix elements of two currents [7,8], e.g.

$$\langle D^+ \pi^- | (\bar{c}b)_{V-A} (\bar{d}u)_{V-A} | \bar{B}_d \rangle \to \langle \pi^- | (\bar{d}u)_{V-A} | 0 \rangle \, \langle D^+ | (\bar{c}b)_{V-A} | \bar{B}_d \rangle \,. \tag{2}$$

This assumes that the exchange of "non-factorizable" gluons between the π^- and the $(\bar{B}_d \, D^+)$ system can be neglected if the virtuality of the gluons is below $\mu \sim m_b$. The non-leptonic decay amplitude then reduces to the product of a form factor and a decay constant. This assumption is in general not justified, except in the limit of a large number of colors in some cases. It deprives the amplitude of any physical mechanism that could account for rescattering in the final state. "Non-factorizable" radiative corrections must also exist, because the scale dependence of the two sides of (2) is different. Since such corrections at scales larger than μ are taken into account in deriving the effective weak Hamiltonian, it appears rather arbitrary to leave them out below the scale μ. Various generalizations of the naive factorization approach have been proposed, which include new parameters that account for non-factorizable corrections. In their most general form, these generalizations have nothing to do with the original "factorization" ansatz, but amount to a general parameterization of the matrix elements. Such general parameterizations are exact, but at the price of introducing many unknown parameters and eliminating any theoretical input on strong-interaction dynamics.

The second method used to study non-leptonic decays is the hard-scattering approach, which assumes the dominance of hard gluon exchange. The decay amplitude is then expressed as a convolution of a hard-scattering factor with light-cone wave functions of the participating mesons, in analogy with more familiar applications of this method to hard exclusive reactions involving only light hadrons [9,10]. In many cases, the hard-scattering contribution represents the leading term in an expansion in powers of Λ_{QCD}/Q, where Q denotes the hard scale. However, the short-distance dominance of hard exclusive processes is not enforced kinematically and relies crucially on the properties of hadronic wave functions. There is an important difference between light mesons and heavy mesons in this regard, because the light quark in a heavy meson at rest naturally has a small momentum of order $\Lambda_{\rm QCD}$, while for fast light mesons a configuration with a soft quark is suppressed by the endpoint behavior of the meson wave function. As a consequence, the soft (or Feynman) mechanism is power suppressed for hard exclusive processes involving light mesons, but it is of leading power for heavy-meson decays.

It is clear from this discussion that a satisfactory treatment should take into account soft contributions, but also allow us to compute corrections to naive factorization in a systematic way. It is not at all obvious that such a treatment would result in a predictive framework. We have shown that this does indeed happen for most non-leptonic two-body B decays. Our main conclusion is that "non-factorizable" corrections are dominated by hard gluon exchange, while the soft effects that survive in the heavy-quark limit are confined to the (BM_1) system, where M_1 denotes the meson that picks up the spectator quark in the B meson. This result is expressed as a factorization formula, which is valid up to corrections suppressed by powers of $\Lambda_{\rm QCD}/m_b$. At leading power, non-perturbative contributions are parameterized by the physical form factors for the $B \to M_1$ transition and leading-twist light-cone distribution amplitudes of the mesons. Hard perturbative corrections can be computed systematically in a way similar to the hard-scattering approach. On the other hand, because the $B \to M_1$ transition is parameterized by a form factor, we recover the result of naive factorization at lowest order in α_s.

An important implication of the factorization formula is that strong rescattering phases are either perturbative or power suppressed in $\Lambda_{\rm QCD}/m_b$. It is worth emphasizing that the decoupling of M_2 occurs in the presence of soft interactions in the (BM_1) system. In other words, while strong-interaction effects in the $B \to M_1$ transition are not confined to small transverse distances, the other meson M_2 is predominantly produced as a compact object with small transverse extension. The decoupling of soft effects then follows from "color transparency". The color-transparency argument for exclusive B decays has already been noted in the literature [11,12], but it has never been developed into a factorization formula that could be used to obtain quantitative predictions.

The QCD factorization approach described in [1–3] is general and applies to decays into a heavy and a light meson (such as $B \to D\pi$) as well as to decays into two light mesons (such as $B \to \pi\pi$). Factorization does not hold, however, for decays such as $B \to \pi D$ and $B \to D\bar{D}$, in which the meson that does *not* pick up the spectator quark in the B meson is heavy. For the first part in these lectures, we will focus on the case of $B \to D^{(*)}L$ decays (with L a light meson), for which the factorization formula takes its simplest form, and power counting will be relatively straightforward. Occasionally, we will point out what changes when we consider more complicated decays such as $B \to \pi\pi$. A detailed treatment of these processes can be found in [3,13,14].

2   Statement of the Factorization Formula

In this section we summarize the factorization formula for non-leptonic B decays. We introduce relevant terminology and provide definitions of the hadronic quantities that enter the factorization formula as non-perturbative input parameters.

2.1   The Idea of Factorization

In the context of non-leptonic decays the term "factorization" is usually applied to the approximation of the matrix element of a four-fermion operator by the product of a form factor and a decay constant, as illustrated in (2). Corrections to this approximation are called "non-factorizable". We will refer to this approximation as "naive factorization" and use quotes on "non-factorizable" to avoid confusion with the (much less trivial) meaning of factorization in the context of hard processes in QCD. In the latter case, factorization refers to the separation of long-distance contributions to the process from a short-distance part that depends only on the large scale m_b. The short-distance part can be computed in an expansion in the strong coupling $\alpha_s(m_b)$. The long-distance contributions must be computed non-perturbatively or determined experimentally. The advantage is that these non-perturbative parameters are often simpler in structure than the original quantity, or they are process independent. For example, factorization applied to hard processes in inclusive hadron–hadron collisions requires only parton distributions as non-perturbative inputs. Parton distributions are much simpler objects than the original matrix element with two hadrons in the initial state. On the other hand, factorization applied to the $B \to D$ form factor leads to a non-perturbative object (the "Isgur–Wise function"), which is still a function of the momentum transfer. However, the benefit here is that symmetries relate this function to other form factors. In the case of non-leptonic B decays, the simplification is primarily of the first kind (simpler structure). We call those effects non-factorizable (without quotes) which depend on the long-distance properties of the B meson and both final-state mesons combined.

The factorization properties of non-leptonic decay amplitudes depend on the two-meson final state. We call a meson "light" if its mass m remains finite in the heavy-quark limit. A meson is called "heavy" if its mass scales with m_b in the heavy-quark limit, such that m/m_b stays fixed. In principle, we could still have $m \gg \Lambda_{\rm QCD}$ for a light meson. Charm mesons could be considered as light in this sense. However, unless otherwise mentioned, we assume that m is of order $\Lambda_{\rm QCD}$ for a light meson, and we consider charm mesons as heavy. In evaluating the scaling behavior of the decay amplitudes,

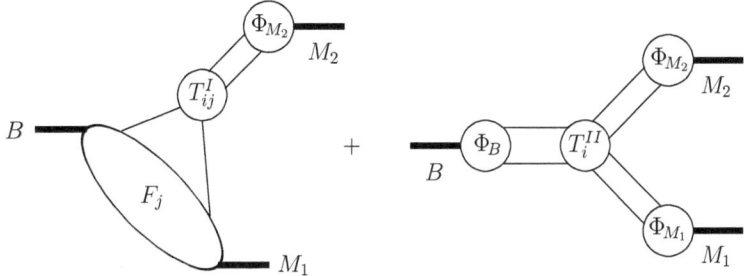

Fig. 1. Graphical representation of the factorization formula. Only one of the two form-factor terms in (3) is shown for simplicity.

we assume that the energies of both final-state mesons (in the B-meson rest frame) scale with m_b in the heavy-quark limit.

2.2   The Factorization Formula

We consider a generic weak decay $B \to M_1 M_2$ in the heavy-quark limit and differentiate between decays into final states containing a heavy and a light meson or two light mesons. Our goal is to show that, up to power corrections of order Λ_{QCD}/m_b, the transition matrix element of an operator O_i in the effective weak Hamiltonian can be written as (a sum over j is understood)

$$\langle M_1 M_2 | O_i | B \rangle = F_j^{B \to M_1}(m_2^2)\, f_{M_2} \int_0^1 du\, T_{ij}^I(u)\, \Phi_{M_2}(u)$$

if M_1 is heavy and M_2 is light,

$$\langle M_1 M_2 | O_i | B \rangle = F_j^{B \to M_1}(m_2^2)\, f_{M_2} \int_0^1 du\, T_{ij}^I(u)\, \Phi_{M_2}(u)\ +\ (M_1 \leftrightarrow M_2)$$

$$+\ f_B f_{M_1} f_{M_2} \int_0^1 d\xi\, du\, dv\, T_i^{II}(\xi, u, v)\, \Phi_B(\xi)\, \Phi_{M_1}(v)\, \Phi_{M_2}(u)$$

if M_1 and M_2 are both light.   (3)

Here $F_j^{B \to M}(m^2)$ denotes a $B \to M$ form factor evaluated at $q^2 = m^2$, $m_{1,2}$ are the light meson masses, and $\Phi_X(u)$ is the light-cone distribution amplitude for the quark–antiquark Fock state of the meson X. These non-perturbative quantities will be defined below. $T_{ij}^I(u)$ and $T_i^{II}(\xi, u, v)$ are hard-scattering functions, which are perturbatively calculable. The factorization formula in its general form is represented graphically in Fig. 1.

The second equation in (3) applies to decays into two light mesons, for which the spectator quark in the B meson (in the following simply referred to as the "spectator quark") can go to either of the final-state mesons. An example is the decay $B^- \to \pi^0 K^-$. If the spectator quark can go only to one of the final-state mesons, as for example in $\bar{B}_d \to \pi^+ K^-$, we call this meson M_1, and the second form-factor term on the right-hand side of (3) is absent. The formula simplifies when the spectator quark goes to a heavy meson (first equation in (3)), such as in $\bar{B}_d \to D^+\pi^-$. Then the second term in Fig. 1, which accounts for hard interactions with the spectator quark, can be dropped because it is power suppressed in the heavy-quark limit. In the opposite situation that the spectator quark goes to a light meson but the other meson is heavy, factorization does not hold, because the heavy meson is neither fast nor small and cannot be factorized from the $B \to M_1$ transition. Finally, notice that annihilation topologies do not appear in the factorization formula, since they do not contribute at leading order in the heavy-quark expansion.

Any hard interaction costs a power of α_s. As a consequence, the hard-spectator term in the second formula in (3) is absent at order α_s^0. Since at this order the functions $T_{ij}^I(u)$ are independent of u, the convolution integral results in the normalization of the meson distribution amplitude, and (3) reproduces naive factorization. The factorization formula allows us to compute radiative corrections to this result to all orders in α_s. Further corrections are suppressed by powers of $\Lambda_{\mathrm{QCD}}/m_b$ in the heavy-quark limit.

The significance and usefulness of the factorization formula stems from the fact that the non-perturbative quantities appearing on the right-hand side of the two equations in (3) are much simpler than the original non-leptonic matrix elements on the left-hand side. This is because they either reflect universal properties of a single meson (light-cone distribution amplitudes) or refer only to a $B \to$ meson transition matrix element of a local current (form factors). While it is extremely difficult, if not impossible [15], to compute the original matrix element $\langle M_1 M_2 | O_i | B \rangle$ in lattice QCD, form factors and light-cone distribution amplitudes are already being computed in this way, although with significant systematic errors at present. Alternatively, form factors can be obtained using data on semi-leptonic decays, and light-cone distribution amplitudes by comparison with other hard exclusive processes.

After having presented the most general form of the factorization formula, we will from now on restrict ourselves to the case of heavy-light final states. Then the (simpler) first formula in (3) applies, and only the first term shown in Fig. 1 is present at leading power.

3   Arguments in Favor of Factorization

In this section we provide the basic power-counting arguments that lead to the factorized structure shown in (3). We do so by analyzing qualitatively the hard, soft and collinear contributions to the simplest Feynman diagrams.

3.1   Preliminaries and Power Counting

For concreteness, we label the charm meson which picks up the spectator quark by $M_1 = D^+$ and assign momentum p' to it. The light meson is labeled

$M_2 = \pi^-$ and assigned momentum $q = E\,n_+$, where E is the pion energy in the B rest frame, and $n_\pm^\mu = (1,0,0,\pm 1)$ are four-vectors on the light-cone. At leading power, we neglect the mass of the light meson.

The simplest diagrams that we can draw for a non-leptonic decay amplitude assign a quark and antiquark to each meson. We choose the quark and antiquark momenta in the pion as (with $\bar{u} \equiv 1 - u$)

$$l_q = uq + l_\perp - \frac{l_\perp^2}{4uE}\, n_-\,, \qquad l_{\bar{q}} = \bar{u}q - l_\perp - \frac{l_\perp^2}{4\bar{u}E}\, n_-\,. \tag{4}$$

Note that $q \neq l_q + l_{\bar{q}}$, but the off-shellness $(l_q + l_{\bar{q}})^2$ is of the same order as the light meson mass, which we can neglect at leading power. A similar decomposition (with longitudinal momentum fraction v and transverse momentum $l_\perp'$) is used for the charm meson.

To prove the factorization formula (3) for the case of heavy-light final states, one has to show that:

i) There is no leading (in powers of $\Lambda_{\mathrm{QCD}}/m_b$) contribution to the amplitude from the endpoint regions $u \sim \Lambda_{\mathrm{QCD}}/m_b$ and $\bar{u} \sim \Lambda_{\mathrm{QCD}}/m_b$.

ii) One can set $l_\perp = 0$ in the amplitude (more generally, expand the amplitude in powers of $l_\perp$) after collinear subtractions, which can be absorbed into the pion wave function. This, together with i), guarantees that the amplitude is legitimately expressed in terms of the light-cone distribution amplitudes of pion.

iii) The leading contribution comes from $\bar{v} \sim \Lambda_{\mathrm{QCD}}/m_b$ (the region where the spectator quark enters the charm meson as a soft parton), which guarantees the absence of a hard spectator interaction term.

iv) After subtraction of infrared contributions corresponding to the light-cone distribution amplitude and the form factor, the leading contributions to the amplitude come only from internal lines with virtuality that scales with m_b.

v) Non-valence Fock states are non-leading.

The requirement that after subtractions virtualities should be large is obvious to guarantee the infrared finiteness of the hard scattering functions T_{ij}^I. Let us comment on setting transverse momenta in the wave functions to zero and on endpoint contributions. Neglecting transverse momenta requires that we count them as order Λ_{QCD} when comparing terms of different magnitude in the scattering amplitude. This conforms to our intuition and the assumption of the parton model, that intrinsic transverse momenta are limited to hadronic scales. However, in QCD transverse momenta are not limited, but logarithmically distributed up to the hard scale. The important point is that contributions that violate the starting assumption of limited transverse momentum can be absorbed into the universal light-cone distribution amplitudes. The statement that transverse momenta can be counted of order Λ_{QCD} is to be understood after these subtractions have been performed.

The second comment concerns endpoint contributions in the convolution integrals over longitudinal momentum fractions. These contributions are dangerous, because we may be able to demonstrate the infrared safety of the hard-scattering amplitude under assumption of generic u and independent of the shape of the meson distribution amplitude, but for $u \to 0$ or $u \to 1$ a propagator that was assumed to be off-shell approaches the mass-shell. If such a contribution were of leading power, we would not expect the perturbative calculation of the hard-scattering functions to be reliable.

Estimating endpoint contributions requires knowledge of the endpoint behavior of the light-cone distribution amplitude. Since it enters the factorization formula at a renormalization scale of order m_b, we can use the asymptotic form $\Phi_X(u) \approx 6u\bar{u}$ to estimate the endpoint contribution. (More generally, we only have to assume that the distribution amplitude at a given scale has the same endpoint behavior as the asymptotic amplitude. This is generally the case, unless there is a conspiracy of terms in the Gegenbauer expansion of the distribution amplitude. If such a conspiracy existed at some scale, it would be destroyed by evolving the distribution amplitude to a different scale.) We count a light-meson distribution amplitude as order Λ_{QCD}/m_b in the endpoint region (defined as the region the quark or antiquark momentum is of order Λ_{QCD}), and order 1 away from the endpoint, i.e. (for $X = P, V_{\parallel}$)

$$\Phi_X(u) \sim \begin{cases} 1\,; & \text{generic } u, \\ \Lambda_{\text{QCD}}/m_b\,; & u,\ \bar{u} \sim \Lambda_{\text{QCD}}/m_b. \end{cases} \qquad (5)$$

Note that the endpoint region has a size of order Λ_{QCD}/m_b, so that the endpoint suppression is $\sim (\Lambda_{\text{QCD}}/m_b)^2$. This suppression has to be weighted against potential enhancements of the partonic amplitude when one of the propagators approaches the mass shell. The counting for B mesons, or heavy mesons in general, is different. Naturally, the heavy quark carries almost all of the meson momentum, and hence we count

$$\Phi_B(\xi) \sim \begin{cases} m_b/\Lambda_{\text{QCD}}\,; & \xi \sim \Lambda_{\text{QCD}}/m_b, \\ 0\,; & \xi \sim 1. \end{cases} \qquad (6)$$

The zero probability for a light spectator with momentum of order m_b must be understood as a boundary condition for the wave function renormalized at a scale much below m_b. There is a small probability for hard fluctuations that transfer large momentum to the spectator. This "hard tail" is generated by evolution of the wave function from a hadronic scale to a scale of order m_b [16,17]. If we assume that the initial distribution at the hadronic scale falls sufficiently rapidly for $\xi \gg \Lambda_{\text{QCD}}/m_b$, this remains true after evolution. We shall assume a sufficiently fast fall-off, so that, for the purposes of power counting, the probability that the spectator-quark momentum is of order m_b can be set to zero. The same counting applies to the D meson. (Despite the fact that the charm meson has momentum of order m_b, we do not need to

Fig. 2. Leading contribution to the hard-scattering kernels $T_{ij}^I(u)$. The weak decay of the b quark through a four-fermion operator is represented by the black square.

distinguish the rest frames of B and D for the purpose of power counting, because the two frames are not connected by a parametrically large boost. In other words, the components of the spectator quark in the D meson are still of order Λ_{QCD}.)

3.2   Non-leptonic Decay Amplitudes

We now turn to a qualitative discussion of the lowest-order and one-gluon exchange diagrams that could contribute to the hard-scattering kernels $T_{ij}^I(u)$ in (3). In the figures which follow, the two lines directed upward represent π^-, the lines on the left represent $\bar{B}_d$, and the lines on the right represent D^+.

Lowest-order diagram: There is a single diagram with no hard gluon interactions shown in Fig. 2. According to (6) the spectator quark is soft, and since it does not undergo a hard interaction it is absorbed as a soft quark by the recoiling meson. This is evidently a contribution to the left-hand diagram of Fig. 1, involving the $B \to D$ form factor. The hard subprocess in Fig. 2 is just given by the insertion of a four-fermion operator, and hence it does not depend on the longitudinal momentum fraction u of the two quarks that form the emitted π^-. Consequently, the lowest-order contribution to $T_{ij}^I(u)$ in (3) is independent of u, and the u-integral reduces to the normalization condition for the pion distribution amplitude. The result is, not surprisingly, that the factorization formula reproduces the result of naive factorization if we neglect gluon exchange. Note that the physical picture underlying this lowest-order process is that the spectator quark (which is part of the $B \to D$ form factor) is soft. If this is the case, the hard-scattering approach misses the leading contribution to the non-leptonic decay amplitude.

Putting together all factors relevant to power counting, we find that in the heavy-quark limit the decay amplitude for a decay into a heavy-light final state (in which the spectator quark is absorbed by the heavy meson) scales as

$$\mathcal{A}(\bar{B}_d \to D^+\pi^-) \sim G_F m_b^2 F^{B\to D}(0) f_\pi \sim G_F m_b^2 \Lambda_{\text{QCD}}. \tag{7}$$

Other contributions must be compared with this scaling rule.

Fig. 3. Diagrams at order α_s that need not be calculated.

Factorizable diagrams: In order to justify naive factorization as the leading term in an expansion in α_s and $\Lambda_{\rm QCD}/m_b$, we must show that radiative corrections are either suppressed in one of these two parameters, or already contained in the definition of the form factor and the pion decay constant. Consider the graphs shown in Fig. 3. The first three diagrams are part of the form factor and do not contribute to the hard-scattering kernels. Since the first and third diagrams contain leading contributions from the region in which the gluon is soft, they should not be considered as corrections to Fig. 2. However, this is of no consequence since these soft contributions are absorbed into the physical form factor.

The fourth diagram in Fig. 3 is also factorizable. In general, this graph would split into a hard contribution and a contribution to the evolution of the pion distribution amplitude. However, as the leading-order diagram in Fig. 2 involves only the normalization integral of the pion distribution amplitude, the sum of the fourth diagram in Fig. 3 and the wave-function renormalization of the quarks in the emitted pion vanishes. In other words, these diagrams would renormalize the $(\bar{u}d)$ light-quark current, which however is conserved.

"Non-factorizable" vertex corrections: We now begin the analysis of "non-factorizable" diagrams, i.e. diagrams containing gluon exchanges that cannot be associated with the $B \to D$ form factor or the pion decay constant. At order α_s, these diagrams can be divided into three groups: vertex corrections, hard spectator interactions, and annihilation diagrams.

The vertex corrections shown in Fig. 4 violate the naive factorization ansatz (2). One of the key observations made in [1,2] is that these diagrams are calculable nonetheless. They form an order-α_s contribution to the hard-scattering kernels $T_{ij}^I(u)$. To demonstrate this, we have to show that: i) The transverse momentum of the quarks that form the pion can be neglected at leading power, i.e. the two momenta in (4) can be approximated by uq and $\bar{u}q$, respectively. This guarantees that only a convolution in the longitudinal momentum fraction u appears in the factorization formula. ii) The contribution from the soft-gluon region and gluons collinear to the direction of the pion is power suppressed. In practice, this means that the sum of these diagrams cannot contain any infrared divergences at leading power in $\Lambda_{\rm QCD}/m_b$.

Neither of the two conditions holds true for any of the four diagrams individually, as each of them separately contains collinear and infrared divergences. However, the infrared divergences cancel when one sums over the gluon attachments to the two quarks comprising the emission pion ((a+b), (c+d) in Fig. 4). This cancellation is a technical manifestation of Bjorken's

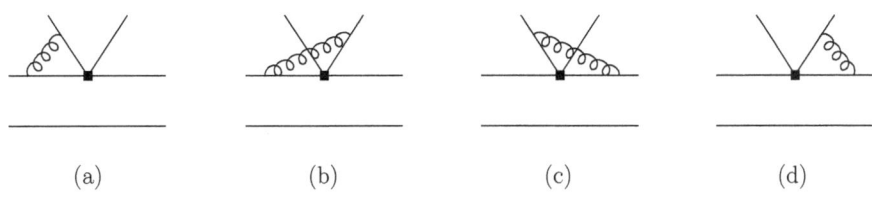

Fig. 4. "Non-factorizable" vertex corrections.

color-transparency argument [11], stating that soft gluon interactions with the emitted color-singlet $(\bar{u}d)$ pair are suppressed because they interact with the color dipole moment of the compact light-quark pair. Collinear divergences cancel after summing over gluon attachments to the b and c quark lines ((a+c), (b+d) in Fig. 4). Thus the sum of the four diagrams (a–d) involves only hard gluon exchange at leading power. Because the hard gluons transfer large momentum to the quarks that form the emission pion, the hard-scattering factor now results in a non-trivial convolution with the pion distribution amplitude. "Non-factorizable" contributions are therefore non-universal, i.e. they depend on the quantum numbers of the final-state mesons.

Note that the color-transparency argument, and hence the cancellation of soft gluon effects, applies only if the $(\bar{u}d)$ pair is compact. This is not the case if the emitted pion is formed in a very asymmetric configuration, in which one of the quarks carries almost all of the pion momentum. Since the probability for forming a pion in such an endpoint configuration is of order $(\Lambda_{\rm QCD}/m_b)^2$, they could become important only if the hard-scattering amplitude favored the production of these asymmetric pairs, i.e. if $T_{ij}^I \sim 1/u^2$ for $u \to 0$ (or $T_{ij}^I \sim 1/\bar{u}^2$ for $u \to 1$). However, such strong endpoint singularities in the hard-scattering amplitude do not occur.

To complete the argument, we have to show that all other types of contributions to the non-leptonic decay amplitudes are power suppressed in the heavy-quark limit. This includes interactions with the spectator quark, weak annihilation graphs, and contributions from higher Fock components of the meson wave functions. This will be done in Sect. 4. In summary, then, for hadronic B decays into a light emitted and a heavy recoiling meson the first factorization formula in (3) holds. At order α_s, the hard-scattering kernels $T_{ij}^I(u)$ are computed from the diagrams shown in Figs. 2 and 4. Naive factorization follows when one neglects all corrections of order $\Lambda_{\rm QCD}/m_b$ and α_s. The factorization formula allows us to compute systematically corrections to higher order in α_s, but still neglects power corrections.

3.3   Remarks on Final-State Interactions

Some of the loop diagrams entering the calculation of the hard-scattering kernels have imaginary parts, which contribute to the strong rescattering phases. It follows from our discussion that these imaginary parts are of order

α_s or $\Lambda_{\mathrm{QCD}}/m_b$. This demonstrates that strong phases vanish in the heavy-quark limit (unless the real parts of the amplitudes are also suppressed). Since this statement goes against the folklore that prevails from the present understanding of this issue, and since the subject of final-state interactions (and of strong-interaction phases in particular) is of paramount importance for the interpretation of CP-violating observables, a few additional remarks are in order.

Final-state interactions are usually discussed in terms of intermediate hadronic states. This is suggested by the unitarity relation (taking $B \to \pi\pi$ for definiteness)

$$\mathrm{Im}\,\mathcal{A}_{B\to\pi\pi} \sim \sum_n \mathcal{A}_{B\to n}\,\mathcal{A}^*_{n\to\pi\pi}\,, \tag{8}$$

where n runs over all hadronic intermediate states. We can also interpret the sum in (8) as extending over intermediate states of partons. The partonic interpretation is justified by the dominance of hard rescattering in the heavy-quark limit. In this limit, the number of physical intermediate states is arbitrarily large. We may then argue on the grounds of parton–hadron duality that their average is described well enough (up to $\Lambda_{\mathrm{QCD}}/m_b$ corrections, say) by a partonic calculation. This is the picture implied by (3). The hadronic language is in principle exact. However, the large number of intermediate states makes it intractable to observe systematic cancellations, which usually occur in an inclusive sum over hadronic intermediate states.

A particular contribution to the right-hand side of (8) is elastic rescattering ($n = \pi\pi$). The energy dependence of the total elastic $\pi\pi$-scattering cross section is governed by soft pomeron behavior. Hence the strong-interaction phase of the $B \to \pi\pi$ amplitude due to elastic rescattering alone increases slowly in the heavy-quark limit [18]. On general grounds, it is rather improbable that elastic rescattering gives an appropriate representation of the imaginary part of the decay amplitude in the heavy-quark limit. This expectation is also borne out in the framework of Regge behavior, as discussed in [18], where the importance (in fact, dominance) of inelastic rescattering was emphasized. However, this discussion left open the possibility of soft rescattering phases that do not vanish in the heavy-quark limit, as well as the possibility of systematic cancellations, for which the Regge approach does not provide an appropriate theoretical framework.

Equation (3) implies that such systematic cancellations *do* occur in the sum over all intermediate states n. It is worth recalling that similar cancellations are not uncommon for hard processes. Consider the example of $e^+e^- \to$ hadrons at large energy q. While the production of any hadronic final state occurs on a time scale of order $1/\Lambda_{\mathrm{QCD}}$ (and would lead to infrared divergences if we attempted to describe it using perturbation theory), the inclusive cross section given by the sum over all hadronic final states is described very well by a $(q\bar{q})$ pair that lives over a short time scale of order $1/q$. In close analogy, while each particular hadronic intermediate state

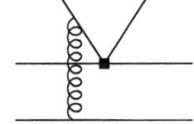 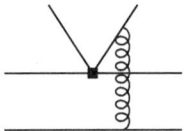

Fig. 5. "Non-factorizable" spectator interactions.

n in (8) cannot be described using a partonic language, the sum over all intermediate states is accurately represented by a $(q\bar{q})$ fluctuation of small transverse size of order $1/m_b$. Because the $(q\bar{q})$ pair is small, the physical picture of rescattering is very different from elastic $\pi\pi$ scattering.

In perturbation theory, the pomeron is associated with two-gluon exchange. The analysis of two-loop contributions to the non-leptonic decay amplitude in [2] shows that the soft and collinear cancellations that guarantee the partonic interpretation of rescattering extend to two-gluon exchange. Hence, the soft final-state interactions are again subleading as required by the validity of (3). As far as the hard rescattering contributions are concerned, two-gluon exchange plus ladder graphs between a compact $(q\bar{q})$ pair with energy of order m_b and transverse size of order $1/m_b$ and the other pion does not lead to large logarithms, and hence there is no possibility to construct the (hard) pomeron. Note the difference with elastic vector-meson production through a virtual photon, which also involves a compact $(q\bar{q})$ pair. However, in this case one considers $s \gg Q^2$, where $\sqrt{s}$ is the photon–proton center-of-mass energy and Q the virtuality of the photon. This implies that the $(q\bar{q})$ fluctuation is born long before it hits the proton. It is this difference of time scales, non-existent in non-leptonic B decays, that permits pomeron exchange in elastic vector-meson production in γ^*p collisions.

4   Power-Suppressed Contributions

Up to this point we have presented arguments in favor of factorization of non-leptonic B-decay amplitudes in the heavy-quark limit. It is now time to show that other contributions not considered so far are indeed power suppressed. This is necessary to fully establish the factorization formula. Besides, it will also provide some estimates of power corrections to the heavy-quark limit.

We start by discussing interactions involving the spectator quark and weak annihilation contributions, before turning to the more delicate question of the importance of non-valence Fock states.

4.1   Interactions with the Spectator Quark

Clearly, the diagrams shown in Fig. 5 cannot be associated with the form-factor term in the factorization formula (3). We will now show that for B decays into a heavy-light final state their contribution is power suppressed in the heavy-quark limit. (This suppression does *not* occur for decays into two

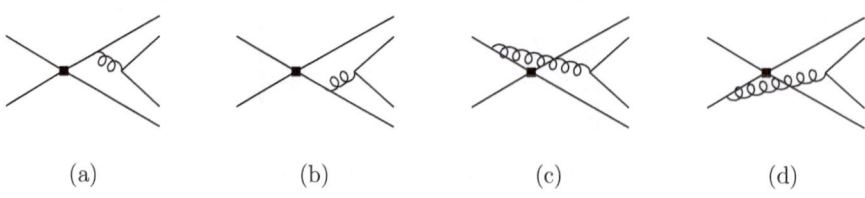

(a)   (b)   (c)   (d)

Fig. 6. Annihilation diagrams.

light mesons, where hard spectator interactions contribute at leading power. In this case, they contribute to the kernels T_i^{II} in the factorization formula (second term in Fig. 1).)

In general, "non-factorizable" diagrams involving an interaction with the spectator quark would impede factorization if there existed a soft contribution at leading power. While such terms are present in each of the two diagrams separately, they cancel in the sum over the two gluon attachments to the $(\bar{u}d)$ pair by virtue of the same color-transparency argument that was applied to the "non-factorizable" vertex corrections.

Focusing again on decays into a heavy and a light meson, such as $\bar{B}_d \to D^+\pi^-$, we still need to show that the contribution remaining after the soft cancellation is power suppressed relative to the leading-order contribution (7). A straightforward calculation leads to the following (simplified) result for the sum of the two diagrams:

$$
\mathcal{A}(\bar{B}_d \to D^+\pi^-)_{\mathrm{spec}} \sim G_F\, f_\pi f_D f_B\, \alpha_s
$$
$$
\times \int_0^1 \frac{d\xi}{\xi}\, \Phi_B(\xi) \int_0^1 \frac{d\eta}{\eta}\, \Phi_D(\eta) \int_0^1 \frac{du}{u}\, \Phi_\pi(u)
$$
$$
\sim G_F\, \alpha_s\, m_b\, \Lambda_{\mathrm{QCD}}^2 . \tag{9}
$$

This is indeed power suppressed relative to (7). Note that the gluon virtuality is of order $\xi\eta\, m_b^2 \sim \Lambda_{\mathrm{QCD}}^2$ and so, strictly speaking, the calculation in terms of light-cone distribution amplitudes cannot be justified. Nevertheless, we use (9) to deduce the scaling behavior of the soft contribution.

4.2   Annihilation Topologies

Our next concern are the annihilation diagrams shown in Fig. 6, which also contribute to the decay $\bar{B}_d \to D^+\pi^-$. The hard part of these diagrams could, in principle, be absorbed into hard-scattering kernels of the type T_i^{II}. The soft part, if unsuppressed, would violate factorization. However, we will see that the hard part as well as the soft part are suppressed by at least one power of $\Lambda_{\mathrm{QCD}}/m_b$.

The argument goes as follows. We write the annihilation amplitude as

$$\mathcal{A}(\bar{B}_d \to D^+\pi^-)_{\text{ann}} \sim G_F\, f_\pi f_D f_B\, \alpha_s$$
$$\times \int_0^1 d\xi\, d\eta\, du\, \Phi_B(\xi)\, \Phi_D(\eta)\, \Phi_\pi(u)\, T^{\text{ann}}(\xi, \eta, u)\,, \quad (10)$$

where the dimensionless function $T^{\text{ann}}(\xi, \eta, u)$ is a product of propagators and vertices. The product of decay constants scales as $\Lambda_{\text{QCD}}^4/m_b$. Since $d\xi\, \Phi_B(\xi)$ scales as 1 and so does $d\eta\, \Phi_D(\eta)$, while $du\, \Phi_\pi(u)$ is never larger than 1, the amplitude can only compete with the leading-order result (7) if $T^{\text{ann}}(\xi, \eta, u)$ can be made of order $(m_b/\Lambda_{\text{QCD}})^3$ or larger. Since $T^{\text{ann}}(\xi, \eta, u)$ contains only two propagators, this can be achieved only if both quarks the gluon splits into are soft, in which case $T^{\text{ann}}(\xi, \eta, u) \sim (m_b/\Lambda_{\text{QCD}})^4$. But then $du\, \Phi_\pi(u) \sim (\Lambda_{\text{QCD}}/m_b)^2$, so that this contribution is power suppressed.

4.3 Non-leading Fock States

Our discussion so far concentrated on contributions related to the quark–antiquark components of the meson wave functions. We now present qualitative arguments that justify this restriction to the valence-quark Fock components. Some of these arguments are standard [9,10]. We will argue that higher Fock states yield only subleading contributions in the heavy-quark limit.

Additional hard partons: An example of a diagram that would contribute to a hard-scattering function involving quark–antiquark–gluon components of the emitted meson and the B meson is shown in Fig. 7. For light mesons, higher Fock components are related to higher-order terms in the collinear expansion, including the effects of intrinsic transverse momentum and off-shellness of the partons by gauge invariance. The assumption is that the additional partons are collinear and carry a finite fraction of the meson momentum in the heavy-quark limit. Under this assumption, it is easy to see that adding additional partons to the Fock state increases the number of off-shell propagators in a given diagram (compare Fig. 7 to Fig. 2). This implies power suppression in the heavy-quark expansion. Additional partons in the B-meson wave function are always soft, as is the spectator quark. Nevertheless, when these partons

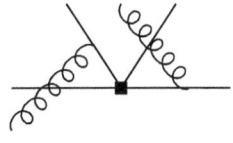

Fig. 7. Contribution to the hard-scattering kernel involving a quark–antiquark–gluon distribution amplitude of the B meson and the emitted light meson.

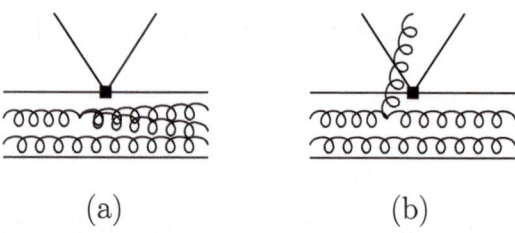

Fig. 8. (a) Soft overlap contribution which is part of the $B \to D$ form factor. (b) Soft overlap with the pion which would violate factorization, if it were unsuppressed.

are connected to the hard-scattering amplitudes the virtuality of the additional propagators is still of order $m_b \Lambda_{\mathrm{QCD}}$, which is sufficient to guarantee power suppression.

Additional soft partons: A more precarious situation may arise when the additional Fock components carry only a small fraction of the meson momentum, contrary to the assumption made above. It is usually argued [9,10] that these configurations are suppressed, because they occupy only a small fraction of the available phase space (since $\int du_i \sim \Lambda_{\mathrm{QCD}}/m_b$ when the parton that carries momentum fraction u_i is soft). This argument does not apply when the process involves heavy mesons. Consider, for example, the diagram shown in Fig. 8 (a) for the decay $B \to D\pi$. Its contribution involves the overlap of the B-meson wave function involving additional soft gluons with the wave function of the D meson, also containing soft gluons. There is no reason to suppose that this overlap is suppressed relative to the soft overlap of the valence-quark wave functions. It represents (part of) the overlap of the "soft cloud" around the b quark with (part of) the "soft cloud" around the c quark after the weak decay. The partonic decomposition of this cloud is unrestricted up to global quantum numbers. (In the case where the B meson decays into two light mesons, there is a form-factor suppression $\sim (\Lambda_{\mathrm{QCD}}/m_b)^{3/2}$ for the overlap of the valence-quark wave functions, but once this price is paid there is again no reason for further suppression of additional soft gluons in the overlap of the B-meson wave function and the wave function of the recoiling meson.)

The previous paragraph essentially repeated our earlier argument against the hard-scattering approach, and in favor of using the $B \to D$ form factor as an input to the factorization formula. However, given the presence of additional soft partons in the $B \to D$ transition, we must now argue that it is unlikely that the emitted pion drags with it one of these soft partons, for instance a soft gluon that goes into the pion wave function, as shown in Fig. 8 (b). Notice that if the $(q\bar{q})$ pair is produced in a color-octet state, at least one gluon (or a further $(q\bar{q})$ pair) must be pulled into the emitted meson if the decay is to result in a two-body final state. What suppresses the process

Fig. 9. Quark–antiquark–gluon distribution amplitude in the endpoint region.

shown in Fig. 8 (b) relative to the one in Fig. 8 (a) even if the emitted $(q\bar{q})$ pair is in a color-octet state?

It is once more color transparency that saves us. The dominant configuration has both quarks carry a large fraction of the pion momentum, and only the gluon might be soft. In this situation we can apply a non-local "operator product expansion" to determine the coupling of the soft gluon to the small $(q\bar{q})$ pair [2]. The gluon endpoint behavior of the $q\bar{q}g$ wave function is then determined by the sum of the two diagrams shown on the right-hand side in Fig. 9. The leading term (for small gluon momentum) cancels in the sum of the two diagrams, because the meson (represented by the black bar) is a color singlet. This cancellation, which is exactly the same cancellation needed to demonstrate that "non-factorizable" vertex corrections are dominated by hard gluons, provides one factor of $\Lambda_{\mathrm{QCD}}/m_b$ needed to show that Fig. 8 (b) is power suppressed relative to Fig. 8 (a).

In summary, we have (qualitatively) covered all possibilities for non-valence contributions to the decay amplitude and find that they are all power suppressed in the heavy-quark limit.

5   Limitations of the Factorization Approach

The factorization formula (3) holds in the heavy-quark limit $m_b \to \infty$. Corrections to the asymptotic limit are power-suppressed in the ratio $\Lambda_{\mathrm{QCD}}/m_b$ and, generally speaking, do not assume a factorized form. Since m_b is fixed to about $5\,\mathrm{GeV}$ in the real world, one may worry about the magnitude of power corrections to hadronic B-decay amplitudes. Naive dimensional analysis would suggest that these corrections should be of order 10% or so. We now discuss several reasons why some power corrections could turn out to be numerically larger than suggested by the parametric suppression factor $\Lambda_{\mathrm{QCD}}/m_b$. Most of these "dangerous" corrections occur in more complicated, rare hadronic B decays into two light mesons, but are absent in decays such as $B \to D\pi$.

5.1   Several Small Parameters

Large non-factorizable power corrections may arise if the leading-power, factorizable term is somehow suppressed. There are several possibilities for such a suppression, given a variety of small parameters that may enter into the non-leptonic decay amplitudes.

i) The hard, "non-factorizable" effects computed using the factorization formula occur at order α_s. Some other interesting effects such as final-state interactions appear first at this order. For instance, strong-interaction phases due to hard interactions are of order α_s, while soft rescattering phases are of order Λ_{QCD}/m_b. Since for realistic B mesons α_s is not particularly large compared to Λ_{QCD}/m_b, we should not expect that these phases can be calculated with great precision. In practice, however, it is probably more important to know that the strong-interaction phases are parametrically suppressed in the heavy-quark limit and thus should be small. (This does not apply if the real part of the decay amplitudes is suppressed for some reason; see below.)

ii) If the leading, lowest-order (in α_s) contribution to the decay amplitude is color suppressed, as occurs for the class-II decay $\bar{B}_d \to \pi^0\pi^0$, then perturbative and power corrections can be sizable. In such a case even the hard strong-interaction phase of the amplitude can be large [1,2]. But at the same time soft contributions could be potentially important, so that in some cases only an order-of-magnitude estimate of the amplitude may be possible.

iii) The effective Hamiltonian (1) contains many Wilson coefficients C_i that are small relative to $C_1 \approx 1$. There are decays for which the entire leading-power contribution is suppressed by small Wilson coefficients, but some power-suppressed effects are not. An example of this type is $B^- \to K^- K^0$. This decay proceeds through a penguin operator $b \to ds\bar{s}$ at leading power. But the annihilation contribution, which is power suppressed, can occur through the current–current operator with large Wilson coefficient C_1. Our approach does not apply to such (presumably) annihilation-dominated decays, unless a systematic treatment of annihilation amplitudes can be found.

iv) Some amplitudes may be suppressed by a combination of small CKM matrix elements. For example, $B \to \pi K$ decays receive large penguin contributions despite their small Wilson coefficients, because the so-called tree amplitude is CKM suppressed. This is not a problem for factorization, since it applies to the penguin and the tree amplitudes. We are not aware of any case (for ordinary B mesons) in which a purely power-suppressed term is CKM enhanced and which would therefore dominate the decay amplitude. (But this situation could occur for $B_c^- \to \bar{D}^0 K^-$, where the QCD dynamics is similar if we consider the charm quark as a light quark.)

5.2 Power Corrections Enhanced by Small Quark Masses

There is another enhancement of power-suppressed effects for some decays into two light mesons, connected with the curious numerical fact that

$$2\mu_\pi \equiv \frac{2m_\pi^2}{m_u + m_d} = -\frac{4\langle \bar{q}q \rangle}{f_\pi^2} \approx 3\,\text{GeV} \tag{11}$$

is much larger than its naive scaling estimate $\Lambda_{\rm QCD}$. (Here $\langle \bar{q}q \rangle = \langle 0|\bar{u}u|0 \rangle = \langle 0|\bar{d}d|0 \rangle$ is the quark condensate.) Consider the contribution of the penguin operator $O_6 = (\bar{d}_i b_j)_{V-A}(\bar{u}_j u_i)_{V+A}$ to the $\bar{B}_d \to \pi^+\pi^-$ decay amplitude. The leading-order graph of Fig. 2 results in the expression

$$\langle \pi^+\pi^-|(\bar{d}_i b_j)_{V-A}(\bar{u}_j u_i)_{V+A}|\bar{B}_d \rangle = i m_B^2 \, F_+^{B \to \pi}(0) \, f_\pi \times \frac{2\mu_\pi}{m_b} \, , \qquad (12)$$

which is formally a $\Lambda_{\rm QCD}/m_b$ power correction compared to the corresponding matrix element of a product of two left-handed currents, but numerically large due to (11). We would not have to worry about such terms if they could all be identified and the factorization formula (3) applied to them, since in this case higher-order perturbative corrections would not contain non-factorizing infrared logarithms. However, this is not the case.

After including radiative corrections, the matrix element on the left-hand side of (12) is expressed as a non-trivial convolution with pion light-cone distribution amplitudes. The terms involving μ_π can be related to two-particle twist-3 (rather than leading twist-2) distribution amplitudes, conventionally called $\Phi_p(u)$ and $\Phi_\sigma(u)$. We find that the radiative corrections to the matrix element in (12) do indeed factorize. However, at the same order there appear twist-3 corrections to the hard spectator interaction shown in Fig. 5, and these contributions contain an endpoint divergence (related to the fact that the distribution amplitudes $\Phi_p(u)$ and $\Phi'_\sigma(u)$ do not vanish at the endpoints). In other words, the twist-3 "corrections" to the hard spectator term in the second factorization formula in (3) relative to the "leading" twist-2 contributions are of the form $\alpha_s \times$ logarithmic divergence, which we interpret as being of order 1. The non-factorizing character of the "chirally-enhanced" power corrections can introduce a substantial uncertainty in some decay modes [3]. As in the related situation for the pion form factor [19], one may argue that the endpoint divergence is suppressed by a Sudakov form factor. However, it is likely that when m_b is not large enough to suppress these chirally-enhanced terms, then it is also not large enough to make Sudakov suppression effective.

We stress that the chirally-enhanced terms do not appear in decays into a heavy and a light meson such as $B \to D\pi$, because these decays have no penguin contribution and no contribution from the hard spectator interaction. Hence, the twist-3 light-cone distribution amplitudes responsible for chirally-enhanced power corrections do not enter in the evaluation of the decay amplitudes.

6   QCD Factorization for Charmless Decays

As we have discussed, hadronic weak decays simplify greatly in the heavy-quark limit $m_b \gg \Lambda_{\rm QCD}$. The underlying physics is that a fast-moving light meson produced by a point-like source (the effective weak Hamiltonian) decouples from soft QCD interactions [11,12,20]. A systematic implementation of this color transparency argument is provided by the QCD factorization approach [1,3]. This scheme makes rigorous predictions in the heavy-quark limit. One can hardly overemphasize the importance of controlling non-leptonic decay amplitudes in the heavy-quark limit. While a few years ago reliable calculations of such amplitudes appeared to be out of reach, we are now in a situation where hadronic uncertainties enter only at the level of power corrections suppressed by the heavy b-quark mass.

Note that this does not imply that non-leptonic amplitudes in the heavy-quark limit are perturbative. (In this respect, our approach is more general than the pQCD scheme [21].) Important non-perturbative effects remain, which can be parameterized in terms of meson decay constants, $B \to M$ transition form factors, and meson light-cone distribution amplitudes. These quantities are an input to the factorization formula, ideally taken from experiment. Theoretical expressions for decay amplitudes obtained using the QCD factorization approach are complicated and depend on many input parameters. When discussing the theoretical uncertainties and limitations of this scheme it is important to distinguish between different classes of parameters. In order of phenomenological importance, these are Standard Model parameters (the parameters $\bar{\rho}$, $\bar{\eta}$ of the unitarity triangle, the quark masses m_s, m_c, and the strong coupling α_s), the renormalization scale (μ), hadronic quantities that can (at least in principle) be determined from data (decay constants, transition form factors), and hadronic quantities that can only indirectly be constrained by data (light-cone distribution amplitudes).

The most important question with regard to phenomenological applications of QCD factorization is that about the numerical size of power corrections. While the importance of the heavy-quark limit to the workings of factorization is evident from a comparison of non-factorizable effects seen in kaon, charm and beauty decays [22], and while there is a lot of evidence (from spectroscopy, exclusive semileptonic decays, and various inclusive decays) that power corrections are small at the b-quark scale [23], it is nevertheless important to address the issue of power corrections in a systematic way. Much effort has been devoted in the past few years to the study of power-suppressed effects, which in general violate factorization. The most important power corrections are proportional to the ratios $2m_K^2/(m_s m_b)$ or $2m_\pi^2/(m_q m_b)$ with $q = u, d$, which are inversely proportional to light-quark masses. Such twist-3 corrections make up for a significant portion of the penguin amplitudes in B decays into light pseudoscalar mesons. It is important that these penguin contributions are calculable despite their power suppression and hence can be included reliably [3]. At the same order, there appear logarithmically divergent twist-3 corrections to the leading-twist hard spectator interactions. These corrections are universal, and their effect can be absorbed into a redefinition of a single hadronic parameter λ_B.

Perhaps the largest uncertainty from power corrections is due to weak annihilation contributions, for which both of the valence quarks of the initial B

meson participate in the weak interactions [21,24]. Annihilation amplitudes violate factorization and thus cannot be reliably computed using the QCD factorization approach. Although we find that with default parameter values the annihilation amplitudes are typically small, their effects can become sizable when the large model uncertainties in their estimate are taken into account [3]. Other types of power corrections, such as soft non-factorizable gluon exchange, have been investigated using QCD sum rules [25] and the renormalon calculus [26,27]. No large corrections of this type have been identified.

While it is a conceptual challenge to gain a better control over the leading power corrections to QCD factorization, perhaps using the framework of the soft-collinear effective theory [28–31], it is important that this approach makes many testable predictions. Their comparison with experimental data can teach us a lot about the importance of power-suppressed effects.

7 Testing Factorization in $B \to \pi K, \pi\pi$ Decays

Deriving constraints on the unitarity triangle from charmless hadronic B decays requires controlling the interference of tree and penguin topologies. This means that one must be able to predict not only the magnitudes of these contributions, but also their relative strong-interaction phase. Fortunately, the crucial aspects of such calculations can be tested using experimental data.

The magnitude of the leading $B \to \pi\pi$ tree amplitude can be probed in the decays $B^\pm \to \pi^\pm\pi^0$, which to an excellent approximation do not receive any penguin contributions. The QCD factorization approach makes an absolute prediction for the corresponding branching ratio [3],

$$\mathrm{Br}(B^\pm \to \pi^\pm\pi^0) = \left[5.3^{+0.8}_{-0.4} \,(\text{pars.}) \pm 0.3 \,(\text{power}) \right] \cdot 10^{-6}$$

$$\times \left[\frac{|V_{ub}|}{0.0035} \frac{F_0^{B \to \pi}(0)}{0.28} \right]^2 , \tag{13}$$

which compares well with the experimental result $(4.9 \pm 1.1) \times 10^{-6}$ [32]. The theoretical uncertainties quoted are due to input parameter variations and to the modeling of power corrections. An additional uncertainty comes from the present error on $|V_{ub}|$ and the $B \to \pi$ form factor.

The magnitude of the leading $B \to \pi K$ penguin amplitude can be probed in the decays $B^\pm \to \pi^\pm K^0$, which to an excellent approximation do not receive any tree contributions. Combining it with the measurement of the tree amplitude just described, a tree-to-penguin ratio can be determined via the relation

$$\varepsilon_{\text{exp}} = \left| \frac{T}{P} \right| = \tan\theta_C \, \frac{f_K}{f_\pi} \left[\frac{2\mathrm{Br}(B^\pm \to \pi^\pm\pi^0)}{\mathrm{Br}(B^\pm \to \pi^\pm K^0)} \right]^{1/2} = 0.20 \pm 0.02 . \tag{14}$$

The experimental value of this ratio is in good agreement with the theoretical prediction $\varepsilon_{\text{th}} = 0.23 \pm 0.04\,(\text{pars.}) \pm 0.04\,(\text{power}) \pm 0.05\,(V_{ub})$ [3], which is independent of form factors but proportional to $|V_{ub}/V_{cb}|$. This is a highly non-trivial test of the QCD factorization approach. Recall that when the first measurements of charmless hadronic decays appeared several authors remarked that the penguin amplitudes were much larger than expected based on naive factorization models. We now see that QCD factorization naturally reproduces the correct magnitude of the tree-to-penguin ratio. This observation also shows that there is no need to supplement the QCD factorization predictions in an ad hoc way by adding enhanced phenomenological penguin amplitudes, such as the "non-perturbative charming penguins" introduced in [33].

QCD factorization predicts that most strong-interaction phases in charmless hadronic B decays are parametrically suppressed in the heavy-quark limit, because

$$\sin\phi_{\text{st}} = O[\alpha_s(m_b), \Lambda_{\text{QCD}}/m_b]. \tag{15}$$

This implies small direct CP asymmetries since, e.g., the following approximate relation holds: $A_{\text{CP}}(\pi^+ K^-) \approx -2\,|T/P|\sin\gamma\,\sin\phi_{\text{st}}$. The suppression results as a consequence of systematic cancellations of soft contributions, which are missed in phenomenological models of final-state interactions. In other schemes the strong-interaction phases are predicted to be larger, and therefore larger CP asymmetries are expected. Present data show no evidence for large direct CP asymmetries in charmless decays [32], but the errors are still too large to distinguish between different theoretical predictions. An important exception is the direct CP asymmetry for the decays $B \to \pi^\pm K^\mp$, which is already measured with high precision. The current world average, $A_{\text{CP}}(\pi^+ K^-) = -0.05 \pm 0.05$ [32], implies a rather small value of the corresponding strong-interaction phase, which is consistent with the expectation that this phase be suppressed in the heavy-quark limit. Specifically, for γ in the range between $60°$ and $90°$, I obtain $\phi_{\text{st}} = (8 \pm 10)°$. Simple physical arguments suggest that the relevant strong-interaction phases in the decays $B \to \pi^\pm K^\mp$ and $B^\mp \to \pi^0 K^\mp$ should be very similar [34]. This observation will become important below.

8   Establishing CP Violation in the Bottom Sector

Measurements of $|V_{ub}|$ in semileptonic decays, $|V_{td}|$ in B–$\bar{B}$ mixing, and $\text{Im}(V_{td}^2)$ from CP violation in K–$\bar{K}$ and B–$\bar{B}$ mixing have firmly established the existence of a CP-violating phase in the CKM matrix. The present situation, often referred to as the "standard analysis" of the unitarity triangle, is summarized in Fig. 10. Three comments are in order concerning this analysis:

i) The measurements of CP asymmetries in kaon physics (ϵ_K and ϵ'/ϵ) and B–$\bar{B}$ mixing ($\sin 2\beta$) probe the imaginary part of V_{td} and so establish

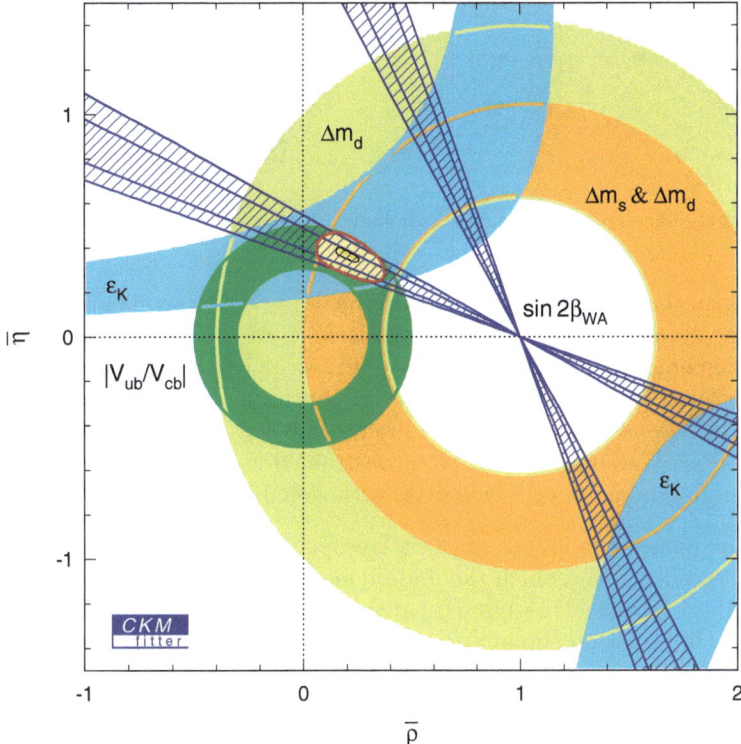

Fig. 10. Standard constraints on the apex $(\bar{\rho}, \bar{\eta})$ of the unitarity triangle [35].

CP violation in the top sector of the CKM matrix.[*] The CKM model predicts that the imaginary part of V_{td} is related, by three-generation unitarity, to the imaginary part of V_{ub}, and that those two elements are (to an excellent approximation) the only sources of CP violation in flavor-changing processes. In order to test this prediction, the next step must be to explore the CP-violating phase $\gamma = \arg(V_{ub}^*)$ in the bottom sector of the CKM matrix.

ii) With the exception of the $\sin 2\beta$ measurement, the standard analysis is limited by large theoretical uncertainties, which dominate the widths of the various bands in the figure. These uncertainties enter via the calculation of hadronic matrix elements. Below, I will discuss some novel methods to constrain the unitarity triangle using charmless hadronic B decays, which are afflicted by smaller hadronic uncertainties and hence provide powerful new tests of the Standard Model, which can complement the standard analysis.

---

[*] Here I adopt the standard phase conventions for the CKM matrix. The corresponding convention-independent statement is that $\mathrm{Im}[(V_{td}V_{ts}^*)/(V_{cd}V_{cs}^*)] \neq 0$ and $\mathrm{Im}[(V_{td}V_{tb}^*)/(V_{cd}V_{cb}^*)] \neq 0$.

Fig. 11. Tree and penguin topologies in charmless hadronic B decays.

iii) With the exception of the measurement of $|V_{ub}|$ in semileptonic B decays, the standard constraints are sensitive to meson–antimeson mixing. Mixing amplitudes are of second order in weak interactions and hence might be most susceptible to effects from physics beyond the Standard Model. The new constraints on $(\bar{\rho}, \bar{\eta})$ discussed below allow a construction of the unitarity triangle that is over-constrained and independent of B–$\bar{B}$ and K–$\bar{K}$ mixing. It is in this sense orthogonal to the standard analysis.

The phase γ can be probed via tree–penguin interference in decays such as $B \to \pi K, \pi\pi$, for which the underlying flavor topologies are illustrated in Fig. 11. Experiment teaches us that amplitude interference is sizable in these decays. Information about γ can be obtained not only from the measurement of direct CP asymmetries ($\sim \sin\gamma$), but also from the study of CP-averaged branching fractions ($\sim \cos\gamma$). The challenge is, of course, to gain theoretical control over the hadronic physics entering the tree-to-penguin ratios in the various decays.

Various ratios of CP-averaged $B \to \pi K, \pi\pi$ branching fractions exhibit a strong dependence on γ and $|V_{ub}|$. It is thus possible to derive constraints on $\bar{\rho}$ and $\bar{\eta}$ from a global analysis of the data in the context of the QCD factorization approach, provided conservative error estimates for power corrections are included. A comprehensive discussion of such an analysis was presented in [3], to which I refer the reader for details. The original result obtained in that paper is reproduced in the upper plot in Fig. 12. It reflects the status of the data as of spring 2001. The lower plot shows an update of this analysis using more recent experimental data [32]. I have also updated two input parameters in order to take into account recent theoretical developments. The new analysis uses $m_s = (100 \pm 25)\,\text{MeV}$ at $\mu = 2\,\text{GeV}$, and $f_B = (200 \pm 30)\,\text{MeV}$. The values adopted in [3] were $m_s = (110 \pm 25)\,\text{MeV}$ and $f_B = (180 \pm 40)\,\text{MeV}$.

The fit is excellent, which $\chi^2 = 0.5$ for three degrees of freedom. There is no problem in accounting for all of the experimental data simultaneously. The inclusion of model estimates of weak annihilation effects enlarges the allowed regions in the $(\bar{\rho}, \bar{\eta})$ plane but is not required to fit the data. Leaving out all annihilation contributions, one still obtains a good fit ($\chi^2 = 0.7$) and similar best-fit values for the Wolfenstein parameters. The comparison of the two plots shown in the figure indicates the effect of the increase in experi-

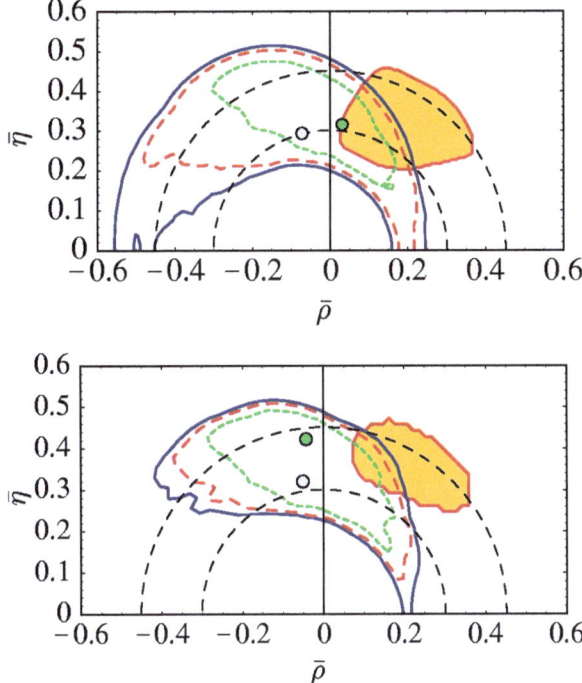

Fig. 12. 95% (solid blue), 90% (dashed red) and 68% (short-dashed green) confidence level contours in the $(\bar{\rho}, \bar{\eta})$ plane obtained from rare hadronic B decays (dark dot = overall best fit; light gray dot = best fit for the default parameter set). The circled region to the right shows the result of the standard CKM fit. Top: status in spring 2001; bottom: update for summer 2002.

mental precision between spring 2001 and summer 2002. The most important conclusion from this analysis is that, with the new data, the combination of results from rare hadronic B decays with the $|V_{ub}|$ measurement in semileptonic decays (dashed circles) excludes $\bar{\eta} = 0$ and so establishes the existence of a CP-violating phase in the bottom sector of the CKM matrix.

The allowed regions obtained from the fit to charmless hadronic decays are compatible with the standard fit (shown by the yellow region), but tend to favor larger γ values. This tendency has been reinforced with the new data. The same trend is seen in an analysis that does not rely on QCD factorization but instead employs general amplitude parameterizations and flavor symmetries [36]. It is tantalizing to speculate about the possible origin of a (still hypothetical) disagreement between the allowed $(\bar{\rho}, \bar{\eta})$ regions obtained from the standard analysis and from charmless hadronic B decays. A conventional explanation of such a discrepancy might be that the errors in lattice calculations of the relevant matrix elements for B_d–$\bar{B}_d$ and B_s–$\bar{B}_s$ mixing have been underestimated. In fact, in a recent paper the value

$\xi = (f_{B_s}\sqrt{B_s})/(f_{B_d}\sqrt{B_d}) = 1.32 \pm 0.10$ was obtained [37], which is significantly larger than the result $\xi = 1.15 \pm 0.05$ used in previous analyses of the unitarity triangle. With such large ξ, values of γ in the vicinity of $90°$ are no longer excluded by the $\Delta m_s/\Delta m_d$ bound.

A more exciting possibility is, of course, to invoke New Physics to explain the discrepancy. Assume first that in charmless hadronic B decays one probes the true value of the CKM phase γ. In this case a discrepancy with the standard analysis would most likely be due to a New Physics contribution to B–$\bar{B}$ mixing. For instance, there could be New Physics affecting B_s–$\bar{B}_s$ mixing. Eliminating the corresponding constraint from the standard analysis one finds that larger values of γ are allowed. This possibility will hopefully soon be checked, when B_s–$\bar{B}_s$ mixing will be explored at the Tevatron. Alternatively, there could be New Physics affecting B_d–$\bar{B}_d$ mixing. In this case one should eliminate the constraints arising from the measurements of Δm_d, $\Delta m_s/\Delta m_d$, and $\sin 2\beta$ from the standard analysis. Then only the constraints from K–$\bar{K}$ mixing and semileptonic B decays remain, which allow for large values of γ. A different possibility would be that the mixing amplitudes are unaffected by New Physics, but that there exist non-standard contributions to $b \to s$ or $b \to d$ FCNC transitions, e.g. from penguin and box diagrams involving the exchange of new heavy particles. (A more exotic model with light SUSY particles has also been considered [38].) In this case, γ measured in $B \to \pi K, \pi\pi$ decays would be a combination of the CKM angle and some new CP-violating phase. Many examples of New Physics models that could yield a significant additional phase have been explored in [39]. A clean test of this possibility would be the measurement of the time-dependent CP asymmetry in $B \to \phi K_S$ decays, which in the Standard Model is due to the interference of a (real) $b \to s\bar{s}s$ penguin amplitude with the B_d–$\bar{B}_d$ mixing amplitude. If there was a New Physics phase $\phi_{\rm NP}$ of the penguin amplitude, then the CP asymmetry in $B \to \phi K_S$ would measure $\sin 2(\beta + \phi_{\rm NP})$, which when compared with the value of $\sin 2\beta$ measured in $B \to J/\psi K_S$ decays would reveal the existence of the phase $\phi_{\rm NP}$ [40]. Note that this strategy would not be invalidated even if there was a New Physics contribution to B_d–$\bar{B}_d$ mixing. In this case β would no longer be given by the CKM phase, but this effect would cancel out in the comparison of the two decay modes.

9   Mixing-Independent Construction of the Unitarity Triangle

Despite of the success of QCD factorization in describing the data, there is an interest in analyzing CKM parameters using methods that rely as little as possible on an underlying theoretical framework. In this last section, I discuss a method for constructing the unitarity triangle from B physics using measurements whose theoretical interpretation is "clean" in the sense that it only relies on assumptions that can be tested using experimental data. I

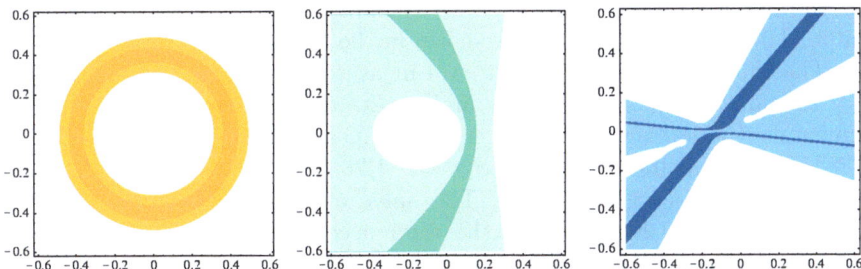

Fig. 13. The three constraints in the $(\bar{\rho}, \bar{\eta})$ plane used in the construction of the CP-b triangle (see text for explanation). Experimental errors are shown at 95% CL. In each plot, the dark band shows the theoretical uncertainty, which is much smaller than the experimental error. This demonstrates the great potential of these methods once the data will become more precise.

call this construction the CP-b triangle, because it probes the existence of a CP-violating phase in the b sector of the CKM matrix. The CP-b triangle is over-determined and can be constructed using already existing data. Most importantly, this construction is insensitive to potential New Physics effects in B–$\bar{B}$ or K–$\bar{K}$ mixing. The present analysis is an update of [41] using the most recent data as of summer 2003.

The first ingredient is the ratio $|V_{ub}/V_{cb}|$ extracted from semileptonic B decays, whose current value is $|V_{ub}/V_{cb}| = 0.09 \pm 0.02$. Several strategies have been proposed to determine $|V_{ub}|$ with an accuracy of about 10% [42–46], which would be a significant improvement. The first plot in Fig. 13 shows the corresponding constraint in the $(\bar{\rho}, \bar{\eta})$ plane. Here and below the narrow, dark-colored band shows the theoretical uncertainty, while the lighter band gives the current experimental value.

The second ingredient is a constraint derived from the ratio of the CP-averaged branching fractions for the decays $B^{\pm} \to \pi^{\pm} K_S$ and $B^{\pm} \to \pi^0 K^{\pm}$, using a generalization of the method suggested in [47]. The experimental inputs to this analysis are a certain tree-to-penguin ratio $\varepsilon_{\mathrm{exp}}$ in (14) and the ratio

$$R_* = \frac{\mathrm{Br}(B^+ \to \pi^+ K^0) + \mathrm{Br}(B^- \to \pi^- \bar{K}^0)}{2[\mathrm{Br}(B^+ \to \pi^0 K^+) + \mathrm{Br}(B^- \to \pi^0 K^-)]} = 0.80 \pm 0.08 \qquad (16)$$

of two CP-averaged $B \to \pi K$ branching fractions [32]. Without any recourse to QCD factorization this method provides a bound on $\cos \gamma$, which can be turned into a determination of $\cos \gamma$ (for fixed value of $|V_{ub}|/V_{cb}|$) when information on the relevant strong-interaction phase ϕ is available. The phase ϕ is bound by experimental data (and very general theoretical arguments) to be small, of order $10°$ [41]. (In the future, this phase can be determined directly from the direct CP asymmetry in $B^{\pm} \to \pi^0 K^{\pm}$ decays.) It is thus conservative to assume that $\cos \phi > 0.8$, corresponding to $|\phi| < 37°$. With this

164 Matthias Neubert

assumption the corresponding allowed region in the $(\bar{\rho}, \bar{\eta})$ plane was analyzed in [3]. The resulting constraint is shown in the second plot in Fig. 13.

The third constraint comes from a measurement of the time-dependent CP asymmetry $S_{\pi\pi} = -\sin 2\alpha_{\text{eff}}$ in $B \to \pi^+\pi^-$ decays. The present experimental situation is still unclear, since the measurements by BaBar ($S_{\pi\pi} = -0.40 \pm 0.22 \pm 0.03$) and Belle ($S_{\pi\pi} = -1.23 \pm 0.41 ^{+0.08}_{-0.07}$) are not in good agreement with each other [48]. The naive average of these results gives $S_{\pi\pi} = -0.58 \pm 0.20$. (Inflating the error according to the PDG prescription would yield $S_{\pi\pi} = -0.58 \pm 0.34$, but for some reason the experimenters usually use the naive error without rescaling, and I will follow their example.) The theoretical expression for the asymmetry is

$$S_{\pi\pi} = -\frac{2\,\text{Im}\,\lambda_{\pi\pi}}{1 + |\lambda_{\pi\pi}|^2}, \quad \text{where} \quad \lambda_{\pi\pi} = e^{-i\phi_d}\,\frac{e^{-i\gamma} + (P/T)_{\pi\pi}}{e^{+i\gamma} + (P/T)_{\pi\pi}}. \tag{17}$$

Here ϕ_d is the CP-violating phase of the B_d–$\bar{B}_d$ mixing amplitude, which in the Standard Model equals 2β. Usually it is argued that for small $(P/T)_{\pi\pi}$ ratio the quantity $\lambda_{\pi\pi}$ is approximately given by $e^{-2i(\beta+\gamma)} = e^{2i\alpha}$, and so apart from a "penguin pollution" the asymmetry $S_{\pi\pi} \approx -\sin 2\alpha$. In order to become insensitive to possible New Physics contributions to the mixing amplitude I adopt a different strategy [3]. I use the measurement $\sin \phi_d = 0.736 \pm 0.049$ [48] and write $e^{-i\phi_d} = \pm(1 - \sin^2\phi_d)^{1/2} - i\sin \phi_d$, with a sign ambiguity in the real part. (The plus sign is suggested by the standard fit of the unitarity triangle.) A measurement of $S_{\pi\pi}$ can then be translated into a constraint on γ (or $\bar{\rho}$ and $\bar{\eta}$), which remains valid even if the $\sin \phi_d$ measurement is affected by New Physics. The result obtained with the current experimental values and assuming $\cos \phi_d > 0$ is shown in the third plot in Fig. 13. The resulting bands for $\cos \phi_d < 0$ are obtained by a reflection about the $\bar{\rho}$ axis. This follows because the expression for $S_{\pi\pi}$ is invariant under the simultaneous replacements $e^{-i\phi_d} \to -e^{i\phi_d}$ and $\gamma \to -\gamma$.

Each of the three constraints in Fig. 13 are, at present, limited by rather large experimental errors, while comparison with Fig. 10 shows that the theoretical limitations are smaller than for the standard analysis. Yet, even at the present level of accuracy it is interesting to combine the three constraints and construct the resulting allowed regions for the apex of the unitarity triangle. The result is shown in the left-hand plot in Fig. 14. Note that the lines corresponding to the new constraints intersect the circles representing the $|V_{ub}|$ constraint at large angles, indicating that the three measurements used in the construction of the CP-b triangle provide highly complementary information on $\bar{\rho}$ and $\bar{\eta}$. There are six (partially overlapping) allowed regions, three corresponding to $\cos \phi_d > 0$ (dark shading) and three to $\cos \phi_d < 0$ (light shading). If we use the information that the measured value of ϵ_K requires a positive value of $\bar{\eta}$, then only the solutions in the upper half-plane remain. Comparison with Fig. 10 shows that one of these regions (corresponding to $\cos \phi_d > 0$) is in perfect agreement with the standard fit. This is highly nontrivial, since with the exception of $|V_{ub}|$ none of the standard constraints are

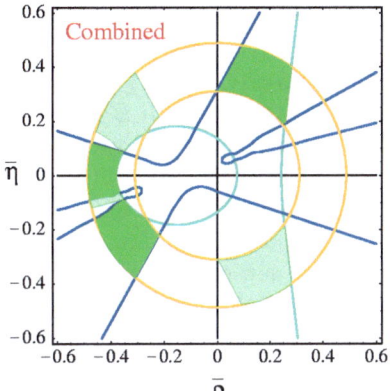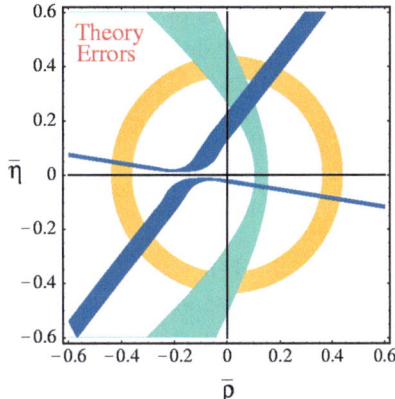

Fig. 14. Left: Allowed regions in the $(\bar{\rho}, \bar{\eta})$ plane obtained from the construction of the CP-b triangle (at 95% CL). The light-shaded areas refer to $\cos\phi_d < 0$. Right: Theoretical error bands for the three constraints combined in the construction of the CP-b triangle.

used in this construction. Interestingly, there is a second allowed region (corresponding to $\cos\phi_d < 0$) which would be consistent with the constraint from ϵ_K but inconsistent with the constraints derived from $\sin 2\beta$ and $\Delta m_s/\Delta m_d$. Such a solution would require a significant New Physics contribution to B–$\bar{B}$ mixing.

10   Outlook

With the recent commissioning of the B factories and the planned emphasis on heavy-flavor physics in future collider experiments, the role of B decays in providing fundamental tests of the Standard Model and potential signatures of new physics will continue to grow. In many cases the principal source of systematic uncertainty is a theoretical one, namely our inability to quantify the non-perturbative QCD effects present in these decays. This is true, in particular, for almost all measurements of CP violation at the B factories.

In these lectures, I have reviewed a rigorous framework for the evaluation of strong-interaction effects for a large class of exclusive, two-body non-leptonic decays of B mesons. The main result is contained in the factorization formula (3), which expresses the amplitudes for these decays in terms of experimentally measurable semi-leptonic form factors, light-cone distribution amplitudes, and hard-scattering functions that are calculable in perturbative QCD. For the first time, therefore, we have a well founded field-theoretic basis for phenomenological studies of exclusive hadronic B decays, and a formal justification for the ideas of factorization.

After the by now precise measurement of $\sin 2\beta$, the study of charmless two-body modes of B mesons is presently the next hottest topic in B physics.

QCD factorization provides the theoretical framework for a systematic analysis of hadronic B decay amplitudes based on the heavy-quark expansion. This theory has already passed successfully several non-trivial tests, and will be tested more thoroughly with more precise data. A global fit to $B \to \pi K, \pi\pi$ decays establishes the existence of a CP-violating phase in the bottom sector of the CKM matrix and tends to favor values of γ near $90°$, somewhat larger than the value suggested by the standard analysis of the unitarity triangle. If this trend were real, it would suggest several possibilities for new flavor physics beyond the Standard Model, ranging from new contributions to B–$\bar{B}$ mixing to non-standard FCNC transitions of the type $b \to sg$ or $b \to s\bar{q}q$. In the future, the construction of the CP-b triangle will provide stringent tests of the Standard Model with small theoretical uncertainties.

Acknowledgment

I would like to thank the organizers of the Schladming Winter School for the invitation to present these lecture, for their hospitality, and for providing a stimulating atmosphere during the school. I am grateful to the students for attending the lectures and contributing with questions and discussions. Finally, I am indebted to my collaborators Martin Beneke, Gerhard Buchalla and Chris Sachrajda, who deserve much credit for these notes. This work was supported by the National Science Foundation under Grant PHY-0098631.

References

1. M. Beneke, G. Buchalla, M. Neubert and C. T. Sachrajda, Phys. Rev. Lett. **83**, 1914 (1999) [hep-ph/9905312].
2. M. Beneke, G. Buchalla, M. Neubert and C. T. Sachrajda, Nucl. Phys. B **591**, 313 (2000) [hep-ph/0006124].
3. M. Beneke, G. Buchalla, M. Neubert and C. T. Sachrajda, Nucl. Phys. B **606**, 245 (2001) [hep-ph/0104110].
4. N. Isgur and M. B. Wise, Phys. Lett. B **237**, 527 (1990).
5. M. A. Shifman and M. B. Voloshin, Sov. J. Nucl. Phys. **45**, 292 (1987) [Yad. Fiz. **45**, 463 (1987)]; Sov. J. Nucl. Phys. **47**, 511 (1988) [Yad. Fiz. **47**, 801 (1988)].
6. G. Buchalla, A. J. Buras and M. E. Lautenbacher, Rev. Mod. Phys. **68**, 1125 (1996) [hep-ph/9512380].
7. D. Fakirov and B. Stech, Nucl. Phys. B **133**, 315 (1978).
8. N. Cabibbo and L. Maiani, Phys. Lett. B **73**, 418 (1978) [Erratum-ibid. B **76**, 663 (1978)].
9. G. P. Lepage and S. J. Brodsky, Phys. Rev. D **22**, 2157 (1980).
10. A. V. Efremov and A. V. Radyushkin, Phys. Lett. B **94**, 245 (1980).
11. J. D. Bjorken, Nucl. Phys. Proc. Suppl. **11**, 325 (1989).
12. M. J. Dugan and B. Grinstein, Phys. Lett. B **255**, 583 (1991).
13. M. Beneke and M. Neubert, Nucl. Phys. B **651**, 225 (2003) [hep-ph/0210085].

14. M. Beneke and M. Neubert, Nucl. Phys. B **675**, 333 (2003) [hep-ph/0308039].
15. L. Maiani and M. Testa, Phys. Lett. B **245**, 585 (1990).
16. A. G. Grozin and M. Neubert, Phys. Rev. D **55**, 272 (1997) [hep-ph/9607366].
17. B. O. Lange and M. Neubert, Phys. Rev. Lett. **91**, 102001 (2003) [hep-ph/0303082].
18. J. F. Donoghue, E. Golowich, A. A. Petrov and J. M. Soares, Phys. Rev. Lett. **77**, 2178 (1996) [hep-ph/9604283].
19. B. V. Geshkenbein and M. V. Terentev, Yad. Fiz. **39**, 873 (1984).
20. H. D. Politzer and M. B. Wise, Phys. Lett. B **257**, 399 (1991).
21. Y. Keum, H. Li and A. I. Sanda, Phys. Lett. B **504**, 6 (2001) [hep-ph/0004004]; Phys. Rev. D **63**, 054008 (2001) [hep-ph/0004173]; Y. Keum and H. Li, Phys. Rev. D **63**, 074006 (2001) [hep-ph/0006001].
22. M. Neubert and A. A. Petrov, Phys. Lett. B **519**, 50 (2001) [hep-ph/0108103].
23. M. Neubert, Phys. Rept. **245**, 259 (1994) [hep-ph/9306320].
24. H. Y. Cheng and K. C. Yang, Phys. Rev. D **64**, 074004 (2001) [hep-ph/0012152].
25. A. Khodjamirian, Nucl. Phys. B **605**, 558 (2001) [hep-ph/0012271].
26. C. N. Burrell and A. R. Williamson, Phys. Rev. D **64**, 034009 (2001) [hep-ph/0101190].
27. T. Becher, M. Neubert and B. D. Pecjak, Nucl. Phys. B **619**, 538 (2001) [hep-ph/0102219]; M. Neubert and B. D. Pecjak, JHEP **0202**, 028 (2002) [hep-ph/0202128].
28. C. W. Bauer, S. Fleming, D. Pirjol and I. W. Stewart, Phys. Rev. D **63**, 114020 (2001) [hep-ph/0011336]; C. W. Bauer, D. Pirjol and I. W. Stewart, Phys. Rev. D **65**, 054022 (2002) [hep-ph/0109045].
29. J. Chay and C. Kim, Phys. Rev. D **65**, 114016 (2002) [hep-ph/0201197].
30. M. Beneke, A. P. Chapovsky, M. Diehl and T. Feldmann, Nucl. Phys. B **643**, 431 (2002) [hep-ph/0206152].
31. R. J. Hill and M. Neubert, Nucl. Phys. B **657**, 229 (2003) [hep-ph/0211018].
32. I use the average of CLEO, BaBar and Belle data as compiled in the talks by R. Bartoldus, P. Dauncey, M. Hazumi, S. Rahatlu and E. Won presented at the Conference on *Flavor Physics and CP Violation*, Philadelphia, USA (May 16–18, 2002), to appear in the proceedings.
33. M. Ciuchini, E. Franco, G. Martinelli and L. Silvestrini, Nucl. Phys. B **501**, 271 (1997) [hep-ph/9703353]; M. Ciuchini, E. Franco, G. Martinelli, M. Pierini and L. Silvestrini, Phys. Lett. B **515**, 33 (2001) [hep-ph/0104126].
34. M. Gronau and J. L. Rosner, Phys. Rev. D **59**, 113002 (1999) [hep-ph/9809384].
35. A. Höcker, H. Lacker, S. Laplace and F. Le Diberder, Eur. Phys. J. C **21**, 225 (2001) [hep-ph/0104062].
36. R. Fleischer and J. Matias, Phys. Rev. D **66**, 054009 (2002) [hep-ph/0204101].
37. A. S. Kronfeld and S. M. Ryan, Phys. Lett. B **543**, 59 (2002) [hep-ph/0206058].
38. T. Becher, S. Braig, M. Neubert and A. L. Kagan, Phys. Lett. B **540**, 278 (2002) [hep-ph/0205274].
39. Y. Grossman, M. Neubert and A. L. Kagan, JHEP **9910**, 029 (1999) [hep-ph/9909297].
40. Y. Grossman and M. P. Worah, Phys. Lett. B **395**, 241 (1997) [hep-ph/9612269].
41. M. Neubert, Nucl. Phys. Proc. Suppl. **121**, 259 (2003) [hep-ph/0207327].
42. M. Neubert, Phys. Rev. D **49**, 4623 (1994) [hep-ph/9312311]; Phys. Lett. B **543**, 269 (2002) [hep-ph/0207002].

43. R. D. Dikeman and N. G. Uraltsev, Nucl. Phys. B **509**, 378 (1998) [hep-ph/9703437];
    I. I. Bigi, R. D. Dikeman and N. Uraltsev, Eur. Phys. J. C **4**, 453 (1998) [hep-ph/9706520].
44. A. F. Falk, Z. Ligeti and M. B. Wise, Phys. Lett. B **406**, 225 (1997) [hep-ph/9705235].
45. C. W. Bauer, Z. Ligeti and M. E. Luke, Phys. Lett. B **479**, 395 (2000) [hep-ph/0002161]; Phys. Rev. D **64**, 113004 (2001) [hep-ph/0107074].
46. M. Neubert, JHEP **0007**, 022 (2000) [hep-ph/0006068];
    M. Neubert and T. Becher, Phys. Lett. B **535**, 127 (2002) [hep-ph/0105217].
47. M. Neubert and J. L. Rosner, Phys. Lett. B **441**, 403 (1998) [hep-ph/9808493];
    Phys. Rev. Lett. **81**, 5076 (1998) [hep-ph/9809311];
    M. Neubert, JHEP **9902**, 014 (1999) [hep-ph/9812396].
48. T. Browder and H. Jawahery, talks presented at the 21th International Symposium on Lepton and Photon Interactions at High Energies, Fermilab, Batavia, Illinois, 11–16 August 2003.

Neutrino Physics – Theory

Walter Grimus

Institute for Theoretical Physics, University of Vienna, Boltzmanngasse 5,
1090 Vienna, Austria, walter.grimus@univie.ac.at

Abstract. We discuss recent developments in neutrino physics and focus, in particular, on neutrino oscillations and matter effects of three light active neutrinos. Moreover, we discuss the difference between Dirac and Majorana neutrinos, neutrinoless $\beta\beta$-decay, absolute neutrino masses and electromagnetic moments. Basic mechanisms and a few models for neutrino masses and mixing are also presented.

1 Motivation

In recent years neutrino physics has gone through a spectacular development (for reviews see, for instance, [1–4]). Data concerning solar and atmospheric neutrino deficits have been accumulated, these deficits have been established as neutrino physics phenomena and the Solar Standard Model [5] has been confirmed. The last steps of this exciting development were the results of the SNO [6] and KamLAND experiments [7]: The SNO experiment provided a model-independent proof of solar $\nu_e \rightarrow \nu_{\mu,\tau}$ transitions and the terrestrial disappearance of $\bar{\nu}_e$ reactor neutrinos in the KamLAND experiment has shown that the puzzle of the solar neutrino deficit is solved by neutrino oscillations [8]. This gives us confidence that the same is true for the atmospheric neutrino deficit as well. For information on the history of neutrino oscillations see [9,10], for the recent experimental history see the contribution of G. Drexlin to these proceedings [4]. General reviews on neutrino physics can be found, for instance, in [11–14].

These lecture notes are motivated by this development and aim at supplying the theoretical background for understanding and assessing it. In view of the importance of neutrino oscillations in this context, we will give a thorough discussion of vacuum neutrino oscillations and matter effects [15,16] (see Sect. 2); the description of the latter is tailored for an understanding of the flavour transformation of solar neutrinos and effects in earth matter. Then, in Sect. 3, we will switch to the subject of the neutrino nature, which is a question of principal interest but has no impact on neutrino oscillations; we will work out the difference between Dirac and Majorana neutrinos and the basics of Majorana neutrino effects. Eventually, in Sect. 4, we will come to the least established field of neutrino physics: models for neutrino masses and mixing. In view of the huge number of models and textures, we cannot try to cover the field but rather discuss a small selection of basic mechanisms

W. Grimus, Neutrino Physics – Theory, Lect. Notes Phys. **629**, 169–214 (2004)
http://www.springerlink.com/ © Springer-Verlag Berlin Heidelberg 2004

for generating neutrino masses and mixing. This selection will necessarily be biased due to personal interest and prejudices. A similar judgement has to be made concerning the literature quoted in these lecture notes; owing to the host of papers which have appeared in recent years only a small selection can be quoted here (for literature on neutrino experiments see [4]). Finally, we present conclusions in Sect. 5.

Abbreviations used in these lecture notes: CC = charged current, NC = neutral current, LBL = long baseline, SBL = short baseline, MSW = Mikheyev-Smirnov-Wolfenstein, LMA = large mixing angle, MM = magnetic moment, EDM = electric dipole moment, SM = Standard Model, SUSY = supersymmetry, GUT = Grand Unified Theory, VEV = vacuum expectation value.

2   Neutrino Oscillations

2.1   Neutrino Oscillations in Vacuum

Here we give a simple and yet quite physical derivation of the neutrino oscillation formula. The first observation is that – as the NC interactions are flavour-blind – neutrino-flavour production and detection proceeds solely via CC interactions with the Hamiltonian density

$$\mathcal{H}_{cc} = \frac{g}{\sqrt{2}} W_\rho^- \sum_{\alpha=e,\mu,\tau} \bar{\ell}_\alpha \gamma^\rho \nu_{\alpha L} + \text{H.c.} \tag{1}$$

The next observation is that flavour transitions are induced by neutrino mixing:

$$\nu_{\alpha L} = \sum_j U_{\alpha j} \nu_{jL} . \tag{2}$$

This relation means that the left-handed flavour fields are not identical with the left-handed components of the neutrino mass eigenfields ν_j corresponding to the mass m_j, but are related via a matrix U, which is determined by the neutrino mass term which will be discussed in Sect. 3. Here we simply assume the existence of the mixing matrix U and the neutrino masses m_j.

We confine ourselves, apart from a few side remarks, to the following basic assumptions:

- There are three active neutrino flavours;
- The mixing matrix U is a 3×3 unitary matrix.

These assumptions are supported by the results of the neutrino oscillation experiments [4].

A crucial observation is that neutrino flavour is defined by the associated charged lepton in production and detection processes; there is no other physical way to define neutrino flavours. Looking at (1) and (2), we find that, in a reasonable approximation, neutrino-flavour states are given by

$$|\nu_\alpha\rangle = \sum_{j=1}^{3} U^*_{\alpha j}|m_j\rangle, \tag{3}$$

where the mass eigenstates $|m_j\rangle$ fulfill the orthogonality condition

$$\langle m_j|m_k\rangle = \delta_{jk}. \tag{4}$$

The state (3) describes a neutrino with flavour α sitting at the position $x = 0$. Every mass eigenstate propagates as a plane wave. If we have a stationary neutrino state with neutrino energy E, we derive from (3) the following form of the propagating state:

$$|\nu_\alpha, x\rangle = \sum_{j=1}^{3} U^*_{\alpha j} e^{-i(Et - p_j x)}|m_j\rangle, \tag{5}$$

with neutrino momenta

$$p_j = \sqrt{E^2 - m_j^2} \simeq E - \frac{m_j^2}{2E} \quad \text{for} \quad E \gg m_j. \tag{6}$$

The latter inequality indicates the relativistic limit of massive neutrinos which is the relevant limit for all neutrino oscillation experiments.

Since the state $|\nu_\alpha, x\rangle$ has flavour α at $x = 0$, we compute the probability to find the flavour β at $x = L$ by $|\langle\nu_\beta|\nu_\alpha, L\rangle|^2$. With the explicit form (5) of $|\nu_\alpha, x\rangle$ and with (6) we readily derive the standard formula of the neutrino transition or survival probability:

$$P_{\nu_\alpha \to \nu_\beta}(L/E) = \left|\sum_{j=1}^{3} U_{\beta j} U^*_{\alpha j} e^{-i m_j^2 L/2E}\right|^2. \tag{7}$$

It is easy to convince oneself that in the case of antineutrinos one simply has to make the replacement $U \to U^*$ in (7).

Looking at (7), we read off the following properties of $P_{\nu_\alpha \to \nu_\beta}$:

- It describes a violation of family lepton numbers;
- It is a function of mass-squared differences, e.g. of $m_2^2 - m_1^2$ and $m_3^2 - m_1^2$;
- It has an oscillatory behaviour in L/E;
- The relation $P_{\nu_\alpha \to \nu_\beta} = P_{\bar\nu_\beta \to \bar\nu_\alpha}$ is fulfilled as a reflection of CPT invariance;
- $P_{\nu_\alpha \to \nu_\beta}$ is invariant under the transformation $U_{\alpha j} \to e^{i\phi_\alpha} U_{\alpha j} e^{i\phi_j}$ with arbitrary phases ϕ_α, ϕ_j.

Specializing (7) to 2-neutrino oscillations with flavours $\alpha \neq \beta$, one obtains the particularly simple formulas

$$U = \begin{pmatrix} \cos\theta & \sin\theta \\ -\sin\theta & \cos\theta \end{pmatrix}, \tag{8}$$

$$P_{\nu_\alpha \to \nu_\beta} = \sin^2 2\theta \times \frac{1}{2}\left(1 - \cos\frac{\Delta m^2 L}{2E}\right), \tag{9}$$

$$P_{\nu_\alpha \to \nu_\beta} = P_{\nu_\beta \to \nu_\alpha}, \tag{10}$$

$$P_{\nu_\alpha \to \nu_\alpha} = P_{\nu_\beta \to \nu_\beta} = 1 - P_{\nu_\alpha \to \nu_\beta}. \tag{11}$$

In (8) the rephasing invariance of $P_{\nu_\alpha \to \nu_\beta}$ has been used to reduce U to a rotation matrix. For 2-neutrino oscillations the probabilities for neutrinos and antineutrinos are the same.

The oscillation phases in (7) have the form

$$\frac{\Delta m^2 L}{2E} = 2.53 \times \left(\frac{\Delta m^2}{1\,\text{eV}^2}\right) \times \left(\frac{1\,\text{MeV}}{E}\right) \times \left(\frac{L}{1\,\text{m}}\right), \tag{12}$$

where Δm^2 is one of the mass-squared differences. Alternatively, the oscillation phase is expressed by the oscillation length L_{osc} through

$$\frac{\Delta m^2 L}{2E} \equiv 2\pi \frac{L}{L_{\text{osc}}} \quad \text{with} \quad L_{\text{osc}} = 2.48\,\text{m} \times \left(\frac{E}{1\,\text{MeV}}\right) \times \left(\frac{1\,\text{eV}^2}{\Delta m^2}\right). \tag{13}$$

In a neutrino oscillation experiment, the sensitivity to Δm^2 is determined by the requirement that the phase (12) is of order one or not much smaller; this requirement depends on the characteristic ratio L/E relevant for the experimental setup. By definition, SBL experiments are characterized by a sensitivity to $\Delta m^2 \gtrsim 0.1$ eV^2, while terrestrial experiments with a sensitivity below 0.1 eV^2 are called LBL experiments. Note that using longer baselines or smaller energies moves the sensitivity to smaller mass-squared differences (see (12)).

No neutrino oscillation experiment has found a mass-squared difference $\Delta m^2 \gtrsim 0.1$ eV^2, apart from the LSND experiment [17]. This experiment, however, has not not been confirmed by any other experiment and requires a forth neutrino which has to be sterile because the invisible Z width at LEP accommodates exactly the three active neutrinos. A *sterile neutrino* is defined as a neutral massive fermion which mixes with the active neutrinos, but has negligible couplings to the W and Z bosons. The result of the LSND experiment is at odds [18] with the combined data sets of either all other SBL experiments or the solar and atmospheric neutrino experiments, when interpreted in terms of 4-neutrino oscillations. See, however, [19], which shows that a 5-neutrino interpretation might reconcile the LSND result with the negative results of the other SBL experiments.

Let us discuss some concrete examples for the sensitivity to Δm^2. For reactor neutrinos the energy is about $E \sim 1$ MeV. Therefore, with $L \sim 1$

km, the sensitivity is $\Delta m^2 \sim 10^{-3}$ eV^2; this is the case of the CHOOZ [20] and Palo Verde [21] experiments. On the other hand, the KamLAND reactor experiment [7] with $L \sim 100$ km is sensitive down to $\Delta m^2 \sim 10^{-5}$ eV^2. Experiments exploiting the source of atmospheric neutrinos with typical energies $E \sim 1$ GeV have as maximal baseline the diameter of the earth, i.e. $L \lesssim 10^4$ km, from where the sensitivity $\Delta m^2 \gtrsim 10^{-4}$ eV^2 follows. Solar neutrinos have a very long baseline of $L \simeq 150 \times 10^6$ km and rather low energies of $E \sim 1 \div 10$ MeV. Thus, in principle with solar neutrino experiments one can reach $\Delta m^2 \sim 10^{-11}$ eV^2.

Quantum-mechanical aspects of neutrino oscillations: Our phenomenological derivation of the oscillation probability (7) needs some quantum-mechanical support (for extensive reviews see [22,23]). The points of our derivation which need justification are the following:

- We have employed stationary neutrino states, i.e., the plane waves have the same energy E but different momenta p_j;
- The usage of plane waves for the neutrino mass eigenstates raises the question how this is compatible with the localization of neutrino source and detection in space.

In order to find limitations of the canonical oscillation probability we have to use a mathematical picture which is as close as possible to the actual situation of an experiment. Neutrinos are never directly observed; thus, for the description of neutrino oscillations it is required to consider the complete neutrino production–detection chain and to use only those quantities or particles which are really observed or manipulated in an experiment [24]. Consequently, we are lead to consider neutrino production and detection as a single big Feynman diagram where both the source and detector particle are described by wave packets which are localized in space, whereas the neutrinos are presented by inner lines of the big diagram; such models are called *external wave packet models* [23,25] and the big diagram is treated in a field-theoretical way [26]. However, due to the macroscopic distance between source and detector the neutrinos are described as on-shell although they are on an inner line of the Feynman diagram [27].

We start with a consideration of the detector particle which is central to the first issue above [28–30]. In all experiments performed so far the detector particle is a bound state with energy E_D, either a nucleon in a nucleus or an electron in an atom. Such states are energy eigenstates. In addition, the detector bound state will experience thermal random movements with a whole spectrum of energies related to the temperature. However, there are no phase correlations between the different energy components of the thermal energy distribution which is, therefore, very well described by a density matrix which is diagonal in energy and the summation over this energy distribution is performed in the *cross section* corresponding to the big production–detection Feynman diagram [30,31]. In other words, in the *amplitude* corresponding to

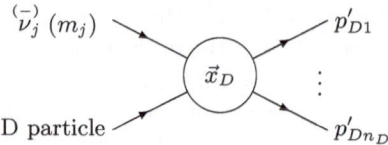

Fig. 1. The particles participating in the detector reaction at $\vec{x}_D$.

the big diagram, we have a definite detector particle energy $E_D + \varepsilon^D_{\text{thermal}}$, with an *incoherent* summation over the distribution of $\varepsilon^D_{\text{thermal}}$ in the cross section. Concerning the n_D final states in the detector reaction (see Fig. 1), we note that energy/momentum measurements are performed and that it is again summed incoherently over these energies E'_{Dk} in the cross section. Thus, looking at Fig. 1 we come to the conclusion that in the *amplitude* the neutrino energy E is fixed and given by

$$E_\nu = \sum_k E'_{Dk} - E_D - \varepsilon^D_{\text{thermal}} . \qquad (14)$$

Since the oscillation probability (7) is directly derived from the cross section, we come to following conclusion which has, in particular, been stressed by H. Lipkin [27–31]:

> Neutrinos with the *same energy* E but different momenta $p_j = \sqrt{E^2 - m_j^2}$ are coherent.

Moreover, summation over the neutrino energies E is effectively performed in the cross section, i.e., it is an incoherent summation and no wave packets are associated with neutrinos of definite mass.

Now we address the justification of the second point. Whereas it is necessary to assume localization in space of the neutrino source and detector wave functions, we stress that the final states in the source and detection reaction can be taken as plane waves; the reason is, as mentioned above, that the measurements performed with them are energy/momentum measurements and for calculating the cross section corresponding to the big Feynman diagram one has to sum *incoherently*, i.e. in the cross section, over the regions in phase space subject to kinematical restrictions according to the experiment. Denoting the detector particle wave function in phase space by $\tilde{\psi}_D(\vec{p})$, a reasonable range for the width of this function is given by $\sigma_D \sim 10^{-3} \div 100$ MeV, where these limits are given for a detector particle bound in an atom or in a nucleus, respectively. In Fig. 2 we have symbolically drawn a detector particle wave function. Looking at this 1-dimensional picture it is easy to find the condition for the existence of oscillations: Assuming for simplicity only two neutrinos, the neutrino mass eigenstates have the same energy E but different momenta p_1, p_2 (6). Suppose that we fix the momenta of the final states of the detector process by p'_{Dk} ($k = 1, \ldots, n_D$, see Fig. 1). By momentum conservation,

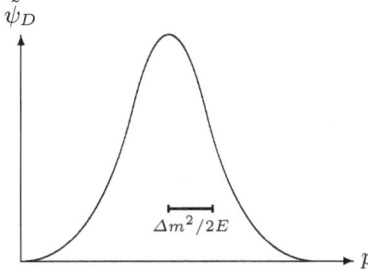

Fig. 2. The wave function of the detector particle. The vertical line indicates the momentum difference $|p_2 - p_1|$.

the values of the detector wave function relevant for the amplitude of the complete production–detection cross section are given by $\tilde{\psi}_D(\sum_k \vec{p}_{Dk} - p_j \vec{\ell})$ for each neutrino mass eigenstate, where $\vec{\ell}$ is a unit vector pointing from the source to the detector particle. Let us assume now that $\sum_k \vec{p}_{Dk} - p_1 \vec{\ell} \simeq 0$. In order to obtain coherence between the neutrino states with mass m_1 and m_2, the momentum p_2 must *not* fulfill $|p_2 - p_1| \gg \sigma_D$, because in that case we have $\tilde{\psi}_D(\sum_k \vec{p}_{Dk} - p_2 \vec{\ell}) \simeq 0$, the two mass eigenstates do not interfere and there are no neutrino oscillations. Therefore, we are lead to the condition

$$|p_2 - p_1| \simeq \Delta m^2/2E \lesssim \sigma_D \tag{15}$$

for neutrino oscillations – see Fig. 2. Denoting the width of the detector wave function in coordinate space by σ_{xD}, then with Heisenberg's uncertainty relation $\sigma_{xD} \sim 1/2\sigma_D$ the condition (15) is rewritten as [27,29,32]

$$\sigma_{xD} \lesssim L_{\text{osc}}/(4\pi) . \tag{16}$$

This condition is trivially fulfilled because σ_{xD} is microscopic whereas L_{osc} is macroscopic! Similar considerations can be made for the neutrino source, with an additional condition taking into account that the source must be unstable [27,29,33]. Thus we can have confidence that the oscillation probability (7) holds for practical purposes.

The present discussion is certainly not the most general one but, as we believe, it is reasonably close to practical applications. It does not include the discussion of a coherence length, i.e. a maximal distance between source and detector until which coherence between different mass eigenstates is maintained. There is a vast literature on that point, see for instance [23,25,26,34,35] and citations therein. One detail can, however, be immediately deduced from our discussion: The better one determines experimentally the energies of all final states in the detection reaction, the better one determines the neutrino energy E. Since the coherence length is approximately obtained by $L_{\text{osc}} \bar{E}/\Delta E$

[34], where $\bar{E}$ is the average neutrino energy and ΔE the neutrino energy spread, it is evident that by pure detector manipulations one can influence the coherence length and, in principle, by infinitely precise measurements of the energies of the final particles in the detector process, one could make it arbitrarily long [29,34].

Some steps in the above investigation of the validity of the oscillation probability (7) have been criticized in the literature – see, for instance, [23,25,35] – and even the external wave packet model is not undisputed [35]. This suggests further investigations into the validity conditions of (7). However, since all attempts up to now to find limitations of (7) accessible to experiment were in vain, the standard formula for the probability of neutrino oscillations in vacuum seems to be very robust.

2.2   Matter Effects

Matter effects [15,16,36] play a very important role in neutrino oscillations. Here we confine ourselves to the case of ordinary matter, which is non-relativistic, electrically neutral and without preferred spin orientation. It consists of electrons, protons and neutrons, i.e. fermions $f = e^-, p, n$. We denote the corresponding matter (number) densities by $N_e = N_p$ and N_n.

We want to present a simple heuristic and straightforward derivation of matter effects. First we notice that the vacuum oscillation probability (7) can be derived by using the Hamiltonian

$$H_\nu^{\text{vac}} = \frac{1}{2E} U \hat{m}^2 U^\dagger , \tag{17}$$

where $\hat{m}$ is the diagonal matrix of neutrino masses. In order to obtain the effective Hamiltonian in matter, we have to add to H_ν^{vac} a term describing matter effects. With the above properties of ordinary matter we have the expectation values

$$\langle \bar{f}_L \gamma_\mu f_L \rangle_{\text{matter}} = \frac{1}{2} N_f \delta_{\mu 0} . \tag{18}$$

Adding CC and NC contributions for neutrinos in the background of ordinary matter, the SM of electroweak interactions together with (18) provides us with the Hamiltonian density

$$\mathcal{H}^{\text{mat}} = \frac{G_F}{\sqrt{2}} \sum_{\alpha = e, \mu, \tau} \nu_\alpha^\dagger (1 - \gamma_5) \nu_\alpha$$
$$\times \sum_f N_f (\delta_{\alpha f} + T_{3f_L} - 2 \sin^2 \theta_W Q_f), \tag{19}$$

where $\delta_{\alpha f}$ comes from a Fierz transformation of the CC interaction,[*] θ_W is the weak mixing angle, and T_{3f_L} and Q_f are the weak isospin and electric

---

[*] This term is non-zero only for ν_e interacting with the background electrons.

charge of the fermion f, respectively. Thus we arrive at the effective flavour Hamiltonian [15,16]

$$H_\nu^{\text{mat}} = \frac{1}{2E} U \hat{m}^2 U^\dagger + \sqrt{2}\, G_F \, \text{diag} \left(N_e - \frac{1}{2} N_n, -\frac{1}{2} N_n, -\frac{1}{2} N_n \right). \qquad (20)$$

Then the evolution of a neutrino state undergoing oscillations and matter effects is calculated by solving

$$i \frac{da}{dx} = H_\nu^{\text{mat}}\, a \quad \text{with} \quad a = \begin{pmatrix} a_e \\ a_\mu \\ a_\tau \end{pmatrix}, \qquad (21)$$

where a_e, a_μ and a_τ are the amplitudes of electron, muon and tau neutrino flavours.

For active neutrinos (ν_α with $\alpha = e, \mu, \tau$) the neutron density N_n can be dropped in the effective Hamiltonian, since it does not induce flavour transitions. Thus for active neutrinos only the electron density is relevant and there are no matter effects in 2-neutrino $\nu_\mu \to \nu_\tau$ transitions. If sterile neutrinos exist, N_n does affect $\nu_\alpha \to \nu_s$ because sterile neutrinos do not experience matter effects, i.e., $N(\nu_s) = 0$ in H_ν^{mat}.

If we consider 2-neutrino oscillations, we obtain from (20) the well-known 2-flavour Hamiltonian by subtraction of an irrelevant diagonal matrix and by using the 2×2 mixing matrix (8) [37]:

$$H_\nu^{\text{mat}} = \frac{1}{4E} \begin{pmatrix} A - \Delta m^2 \cos 2\theta & \Delta m^2 \sin 2\theta \\ \Delta m^2 \sin 2\theta & -A + \Delta m^2 \cos 2\theta \end{pmatrix} \qquad (22)$$

with

$$A = 2\sqrt{2}\, G_F E \left[N(\nu_\alpha) - N(\nu_\beta) \right] \qquad (23)$$

and $N(\nu_e) = N_e - N_n/2$, $N(\nu_\mu) = N(\nu_\tau) = -N_n/2$, $N(\nu_s) = 0$. This equation shows once more what we have discussed above. From now on we do not discuss sterile neutrinos anymore in these lecture notes.

Performing all the analogous procedures for antineutrinos, we obtain $H_{\bar\nu}^{\text{mat}}$ by making the replacements $U \to U^*$, matter term $\to$ $-$matter term in H_ν^{mat} of (20).

In order to have an idea of the strength of matter effects, we estimate the "matter potential" A for the two most important cases, the sun and the earth:

$$A = 2\sqrt{2} G_F E N_e \simeq \begin{cases} \text{core of the sun:} \\ 1.5 \times 10^{-5} \text{ eV}^2 \left(\frac{E}{1\,\text{MeV}} \right), \\ \text{earth matter:} \\ 2.3 \times 10^{-7} \text{ eV}^2 \left(\frac{\rho}{3\,\text{g cm}^{-3}} \right) \left(\frac{E}{1\,\text{MeV}} \right). \end{cases} \qquad (24)$$

The matter potential is the quantity which has to be compared with Δm^2 in order to estimate the influence of matter on neutrino oscillations.

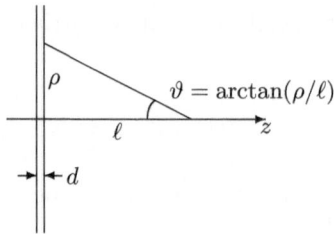

Fig. 3. Scattering at a slab of matter of thickness d of an incident wave propagating in z-direction

In the evolution equation (21) the Hamiltonian is in general a function of the space coordinates through the electron density $N_e(x)$. Looking at the 2-neutrino Hamiltonian (22), a MSW resonance [16] occurs if there is a co-ordinate x_{res} such that

$$A(x_{\text{res}}) = \Delta m^2 \cos 2\theta \tag{25}$$

holds. In this case, the probability of flavour transitions can be very big, in particular, if the neutrino goes adiabatically through the resonance point, as we shall see in the next subsection.

In the following, we will adopt – without loss of generality – the convention $\Delta m^2 > 0$ and $0° \leq \theta \leq 90°$ in the Hamiltonian (22). Therefore, the resonance condition is fulfilled for $\theta < 45°$ and a suitable electron density.

On the validity of the effective flavour Hamiltonian: The heuristic derivation of matter effects resulting in the Hamiltonian (20) did not allow to see under which conditions (20) and (21) hold. We now want to discuss a method which gives us some insight into the limitations of these equations. We follow [38,39]. Suppose we have an incident wave propagating in z-direction which traverses a slab of matter of "infinitesimal" thickness d. Then the incident wave undergoes multiple scattering at the scatterers f with number density N_f in the slab. According to Fig. 3 we sum over all scattered waves hitting a point on the z-axis at a distance ℓ from the slab and obtain

$$N_f d \int_0^\infty \frac{e^{iE\sqrt{\rho^2+\ell^2}}}{\sqrt{\rho^2+\ell^2}} \mathcal{A}_{\nu_\alpha f}(E,\vartheta)\, 2\pi\rho\, d\rho =$$

$$2\pi N_f d \left\{ \frac{e^{iE\sqrt{\rho^2+\ell^2}}}{iE} \mathcal{A}_{\nu_\alpha f}(E,\vartheta) \Big|_0^\infty - \right. \tag{26}$$

$$\left. \int_0^\infty d\rho\, \frac{e^{iE\sqrt{\rho^2+\ell^2}}}{iE\ell\,[1+(\ell/\rho)^2]} \frac{\partial \mathcal{A}_{\nu_\alpha f}(E,\vartheta)}{\partial\vartheta} \right\}.$$

In this equation, $\mathcal{A}_{\nu_\alpha f}$ is the scattering amplitude. Equation (26) can be considerably simplified. Firstly, it is reasonable to assume that in the term in

Table 1. Vector coupling constants of the Hamiltonian density (29).

g_V	e^-	p	n
ν_e	$2\sin^2\theta_w + 1/2$	$-2\sin^2\theta_w + 1/2$	$-1/2$
$\nu_{\mu,\tau}$	$2\sin^2\theta_w - 1/2$	$-2\sin^2\theta_w + 1/2$	$-1/2$

the second line of this equation we can drop the limit $\rho \to \infty$ because that contribution averages out. Secondly, for $1/(E\ell) \ll 1$ and a smooth behaviour of the scattering amplitude, we can also drop the term in the third line of (26). Now we go one step further and consider a slab of finite thickness $D = kd$, where d is "infinitesimal" whereas k is a large integer. With the above-mentioned approximations we compute the phase of incident + transmitted wave through the finite slab as

$$\lim_{k\to\infty} \left\{ 1 + \frac{iED}{k} + \frac{2\pi i N_f D \mathcal{A}_{\nu_\alpha f}(E, \vartheta = 0)}{Ek} \right\}^k =$$
$$\exp\left\{ iED \left[1 + 2\pi N_f \mathcal{A}_{\nu_\alpha f}(E, \vartheta = 0)/E^2 \right] \right\}. \tag{27}$$

The expression in the curly brackets of the first line of this equation, where the incident wave is represented by $1 + iED/k \simeq \exp(iED/k)$, corresponds to the "infinitesimal" slab of thickness $d = D/k$ and $\ell = 0$ (see Fig. 3); the power k indicates that the wave propagates through k such "infinitesimal" slabs. This consideration yields the well-known "index of refraction"

$$n(\nu_\alpha) = 1 + \frac{2\pi}{E^2} \sum_f N_f \mathcal{A}_{\nu_\alpha f}(E, \vartheta = 0), \tag{28}$$

where only the forward-scattering amplitude, i.e. $\vartheta = 0$ contributes. In $n(\nu_\alpha)$ we have indicated that different types of scatterers can be distributed in the slab. Furthermore, the derivation of the "index of refraction" is evidently not specific to neutrino scattering.

The connection with neutrino scattering is made by the Hamiltonian density

$$\mathcal{H}_{\nu_\alpha f} = \frac{G_F}{\sqrt{2}} \bar{\nu}_\alpha \gamma_\rho (1 - \gamma_5) \nu_\alpha \, \bar{f} \gamma^\rho \left(g_V^{(\alpha, f)} - g_A^{(\alpha, f)} \gamma_5 \right) f. \tag{29}$$

The coupling constants $g_V^{(\alpha, f)}$ are found in Table 1. The axial-vector coupling constants do not contribute for ordinary matter (see (18)). As mentioned before, in the $\nu_e e^-$ scattering amplitude, there are CC + NC contributions, otherwise only NC interactions contribute. With the weak forward-scattering amplitude [39]

$$\mathcal{A}_{\nu_\alpha f}(E, \vartheta = 0) = \frac{G_F E}{\sqrt{2\pi}} g_V^{(\alpha, f)}, \tag{30}$$

Equation (27) and Table 1, we compute $2\pi \sum_f N_f \mathcal{A}_{\nu_\alpha f}(E, \vartheta = 0)/E$ and obtain exactly the matter terms as in $\mathcal{H}_\nu^{\text{mat}}$ in (20).

Now we use this second derivation of the matter potential as a means to check the validity of the Hamiltonian (20) [39]. We define two lengths: d_{scatt} is the average distance of scatterers in ordinary matter, $d_{\text{var}} \sim N_f(x)/|\frac{dN_f(x)}{dx}|$ is the typical distance of matter density variations. Then we note the following conditions for the validity of (20) and (21):

* $E d_{\text{scatt}} \gg 1 \Rightarrow \lambda_\nu^{\text{de Broglie}} \ll 2\pi d_{\text{scatt}}$;
* $d_{\text{scatt}} \ll d_{\text{var}}$;
* $d_{\text{scatt}} \ll L_{\text{osc}}$.

The 2nd and 3rd condition arise from the requirement of describing the neutrino state evolution by the differential equation (21). They are trivially fulfilled because d_{scatt} is microscopic, whereas d_{var} and L_{osc} are macroscopic. For instance for the sun we have $d_{\text{var}} \sim 10^4$ km.

The first condition arises from the dropping of the term in the third line of (26). The quantity $\lambda_\nu^{\text{de Broglie}} = 2\pi/E$ is the de Broglie wave length of the neutrino. First we have to estimate d_{scatt}. It is easy the convince oneself that to a very good approximation the electron density in matter is given by

$$N_e \simeq Y_e N_A (\rho/1\,g) , \tag{31}$$

where Y_e is the number of electrons per nucleon, ρ is the matter density in units of g/cm^3 and $N_A \simeq 6.022 \times 10^{23}$ is Avogadro's constant. Let us check the first condition for the sun. According to the Solar Standard Model [5], in the solar core one has $\rho \simeq 150$ and $Y_e \simeq 2/3$, from where it follows that $N_e \simeq 100\,N_A$ cm^{-3} and $d_{\text{scatt}} = N_e^{-1/3} \simeq 0.25$ (23). Thus the first condition above is reformulated as $E\,d_{\text{scatt}} \simeq 130\,E/(1\,\text{MeV}) \gg 1$. For solar neutrinos with $E \gtrsim 1$ MeV this condition is fulfilled. For earth matter this condition holds as well, because d_{scatt} is a little larger than in the solar core.

For other derivations of matter effects see [40] and references therein.

Survival probability for solar neutrinos: We want to conclude this subsection by some general considerations concerning solar neutrinos within the framework of 2-neutrino oscillations. A neutrino produced in the solar core traverses first solar matter, then travels through vacuum to the neutrino detector on earth; during the night, the neutrino has to traverse, in addition, some stretch of earth matter. We denote by $P_{e1,2}^S$ the probability for $|\nu_e\rangle \to |\nu_{1,2}\rangle$ in the sun. Furthermore, the probability of $|\nu_{1,2}\rangle \to |\nu_e\rangle$ transitions in earth matter is called $P_{1,2e}^E$. Note that with the 2×2 mixing matrix U of (8) during the day we have

$$P_{1e}^E = \cos^2\theta \quad \text{and} \quad P_{2e}^E = \sin^2\theta , \tag{32}$$

where θ is the solar neutrino mixing angle. Then the survival probability of solar electron neutrinos is written as [41]

$$P_{\nu_e \to \nu_e} = P_{e1}^S P_{1e}^E + P_{e2}^S P_{2e}^E +$$
$$2\sqrt{P_{e1}^S P_{1e}^E P_{e2}^S P_{2e}^E} \cos(\delta_m + \Delta m^2 L_{\text{vac}}/2E). \qquad (33)$$

In this formula, $L_{\text{vac}} \simeq 150 \times 10^6$ km is the distance between the surface of the sun and the surface of the earth along the neutrino trajectory, i.e. $\Delta m^2 L_{\text{vac}}/2E$ is the phase the neutrino acquires in vacuum. The phase acquired in matter is denoted by δ_m.

Equation (33) is completely general. One obvious questions arises: Under which conditions can the interference term in $P_{\nu_e \to \nu_e}$ be dropped? To investigate this point one can make a rough estimate. The size of the neutrino production region is approximately the diameter ℓ_{core} of the solar core. Therefore, for $\Delta m^2 \ell_{\text{core}}/2E \gtrsim 2\pi$ with $\ell_{\text{core}} \sim 10^5$ km, we obtain the condition

$$\Delta m^2/E > 10^{-14} \text{ eV} \quad \Rightarrow \quad \langle \cos(\delta_m + \Delta m^2 L_{\text{vac}}/2E) \rangle_{\text{core}} = 0. \qquad (34)$$

This argument is not completely correct because of the matter effects, but numerical calculations give the same result [41]. On the other hand, experiments cannot measure the neutrino energy E with infinite precision and energy averaging in the vacuum phase occurs. If $(\Delta m^2 L/2E)(\delta E/E) \gtrsim 2\pi$, where δE is the uncertainty in the measurement of E, with an optimistic assumption of $\delta E/E \sim 10^{-2}$ we again arrive at (34).

For the LMA MSW oscillation solution of the solar neutrino puzzle with $\Delta m^2 \sim 10^{-5}$ eV^2, condition (34) is very well fulfilled, and for the LOW solution – named after its "low" mass-squared difference of $\Delta m^2 \sim 10^{-7}$ eV^2 – this condition is still fulfilled [41]. For the allowed regions in the $\tan^2 \theta$–Δm^2 plane of these solutions see Fig. 4. Note that this figure is relevant for the situation after the SNO results but before the KamLAND result which has ruled out the LOW region in the plot. For the recent history of solar neutrino oscillations see [4]. For the LMA MSW solution the resonance condition (25) plays an important role, as we will see in the next subsection.

2.3   Adiabatic Neutrino Evolution in Matter

Now we want to discuss adiabatic neutrino evolution in matter. We focus on solar neutrinos and confine ourselves to 2-neutrino flavours. Solutions of the 2-neutrino differential equation

$$i\frac{d}{dx}\begin{pmatrix} a_e \\ a_x \end{pmatrix} = H_\nu^{\text{mat}} \begin{pmatrix} a_e \\ a_x \end{pmatrix} \qquad (35)$$

can be found by numerical integration and, for instance, the plots of allowed $\tan^2 \theta$–Δm^2 regions of solar neutrino oscillations like Fig. 4 are usually obtained via such numerical solutions, taking into account averaging over the

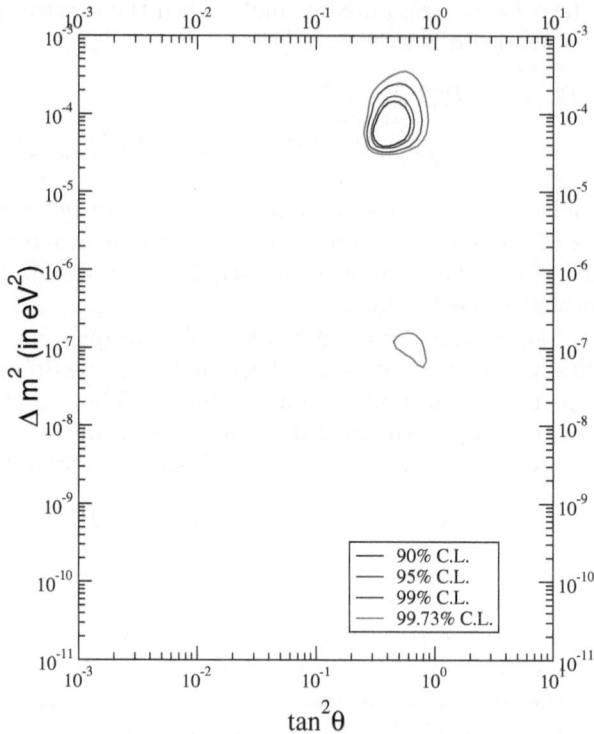

Fig. 4. Allowed regions in the $\tan^2 \theta$–Δm^2 plane for the 2-neutrino oscillation solutions of the solar neutrino deficit. The upper region represents the LMA MSW solution, the lower one the LOW solution. This plot showing the status after the SNO results but before the KamLAND result has been taken from [42].

solar neutrino production region. Equation (35) refers to $\nu_e \to \nu_x$ transitions. In the next subsection, in the context of three neutrino flavours, we will discuss to which ν_μ–ν_τ flavour combination the amplitude a_x refers. The Hamiltonian H_ν^{mat} is given by (22).

In general, solutions of (35) will be non-adiabatic. However, since the result of the KamLAND experiment we know that the solar neutrino puzzle is solved by neutrino oscillations corresponding to the LMW MSW solution; we will argue here that this solution behaves very well adiabatically.

For the consideration of adiabaticity (see, for instance, [43]) we need the eigenvectors of H_ν^{mat}, which are defined $\forall\, x$ by

$$H_\nu^{\mathrm{mat}}(x)\psi_{mj}(x) = E_j(x)\psi_{mj}(x)\,, \tag{36}$$

where

$$\psi_{m1} = \begin{pmatrix} \cos\theta_m \\ -\sin\theta_m \end{pmatrix},\ \psi_{m2} = \begin{pmatrix} \sin\theta_m \\ \cos\theta_m \end{pmatrix}. \tag{37}$$

In the 2-flavour case discussed here and with the real Hamiltonian (22), one parameter, the *matter angle* θ_m, is sufficient to parameterize the two eigenvectors (37). With (22) the matter angle is expressed as

$$\tan 2\theta_m(x) = \frac{\tan 2\theta}{1 - \frac{A(x)}{\Delta m^2 \cos 2\theta}} \quad \text{with} \quad A(x) = 2\sqrt{2}EN_e(x). \tag{38}$$

Note that for $A \to 0$, i.e. negligible matter effects, we obtain $\theta_m \to \theta$, i.e., the matter angle becomes identical with the (vacuum) mixing angle of U of (8).

The full solution of the differential equation (35) can be written as

$$\psi(x) = \sum_{j=1,2} a_j(x)\psi_{mj}(x)e^{-i\varphi_j(x)} \quad \text{with} \quad \varphi_j = \int_{x_0}^{x} dx' E_j(x'). \tag{39}$$

Having in mind solar neutrinos, we identify Δm^2 with the solar mass-squared difference and use the initial condition that at x_0 a ν_e is produced. This leads to

$$\psi(x_0) = \begin{pmatrix} 1 \\ 0 \end{pmatrix} \Rightarrow \begin{cases} a_1(x_0) = \cos\theta_m(x_0), \\ a_2(x_0) = \sin\theta_m(x_0). \end{cases} \tag{40}$$

Equation (39) is completely general. A solution is called adiabatic if the following condition is fulfilled.

$$\text{Adiabaticity}: \quad a_{1,2}(x) \simeq \text{constant.} \tag{41}$$

In that case we have $a_{1,2}(x) \simeq a_{1,2}(x_0)$ and with (40) we obtain

$$P_{e1}^S = \cos^2\theta_m(x_0), \quad P_{e2}^S = \sin^2\theta_m(x_0). \tag{42}$$

Using now that the interference term in (33) can be dropped for the LMA MSW solution and taking into account (32), we compute

$$P_{\nu_e \to \nu_e} = \cos^2\theta_m(x_0)\cos^2\theta + \sin^2\theta_m(x_0)\sin^2\theta$$
$$= \frac{1}{2}\left(1 + \cos 2\theta_m(x_0)\cos 2\theta\right). \tag{43}$$

This is the well-known survival probability for adiabatic neutrino evolution from matter to vacuum. Since we used (32), it does not include earth matter effects.

Let us now consider effects of non-adiabaticity. Plugging the full solution (39) into (35), we derive an equation for the coefficients $a_{1,2}$:

$$\frac{d}{dx}\begin{pmatrix} a_1 \\ a_2 \end{pmatrix} = \begin{pmatrix} 0 & -\frac{d\theta_m}{dx}e^{i\varphi} \\ \frac{d\theta_m}{dx}e^{-i\varphi} & 0 \end{pmatrix}\begin{pmatrix} a_1 \\ a_2 \end{pmatrix}, \tag{44}$$

where we have used the definition

$$\varphi = \varphi_1 - \varphi_2 = \int_{x_0}^{x} dx' \Delta E \quad \text{with} \quad \Delta E = E_1 - E_2. \tag{45}$$

With the Hamiltonian (22) we calculate

$$\Delta E = \frac{1}{2E} \sqrt{(A - \Delta m^2 \cos 2\theta)^2 + (\Delta m^2 \sin 2\theta)^2}. \tag{46}$$

If along the neutrino trajectory the phase factor $\exp(i\varphi)$ in (44) oscillates very often while the change in θ_m is of order one, then $\dot{a}_{1,2}$ will be approximately zero. This suggests the introduction of the "adiabaticity parameter" [44–46]

$$\gamma(x) \equiv \Delta E / 2|\dot{\theta}_m|. \tag{47}$$

Therefore, we conclude

$$\text{Adiabaticity} \Leftrightarrow \gamma \gg 1 \tag{48}$$

along the neutrino trajectory. In vacuum, $\gamma = \infty$, in agreement with $a_{1,2}$ being exactly constant. Non-adiabaticity is quantified by the probability P_c for crossing from ψ_{m1} to ψ_{m2}. It corrects (43) to [45]

$$P_{\nu_e \to \nu_e} = \frac{1}{2} + \left(\frac{1}{2} - P_c\right) \cos 2\theta_m(x_0) \cos 2\theta. \tag{49}$$

Again, as in (43), this form of the survival probability holds where (34) is valid. The adiabaticity parameter γ can be used to find a mathematically exact upper bound on P_c [12].

Reverting to solar neutrinos, we use that the electron density in the sun fulfills

$$N_e(x) \propto \exp(-x/r_0), \tag{50}$$

except for the inner part of the core and toward the edge [5], with $r_0 \simeq R/10.54 \simeq 6.6 \times 10^4$ km $\leftrightarrow 3.3 \times 10^{20}$ MeV^{-1}; R is the solar radius. We want to estimate γ for the case that the neutrino goes through a resonance (25) in a region where the exponential form of the electron density provides a good approximation. Then adiabaticity will be rather well fulfilled if it is fulfilled at the resonance point [44–47]. There, γ is given by

$$\gamma_{\text{res}} = \frac{\Delta m^2 \sin^2 2\theta}{2E \cos 2\theta (|\dot{A}|/A)_{\text{res}}} \simeq \frac{\Delta m^2 r_0}{2E} \times \cos 2\theta \tan^2 2\theta. \tag{51}$$

For large θ and close to the solar edge this estimate for adiabaticity is not correct (see [48]).

Now we turn to the characteristics of the LMA MSW solar neutrino solution with $\Delta m^2 \sim 7 \times 10^{-5}$ eV^2 and $\theta \sim 34°$ [1,2].** Writing the oscillation length (13) as

$$L_{\text{osc}} = 35 \text{ km} \left(\frac{7 \times 10^{-5} \text{ eV}^2}{\Delta m^2}\right) \left(\frac{E}{1 \text{ MeV}}\right), \tag{52}$$

** Before the result of the KamLAND experiment the mass-squared difference was a little lower at $\Delta m^2 \sim 5 \times 10^{-5}$ eV^2 – see, e.g., [41,42].

it is evident that in this case the flavour transition occurs inside the sun. Furthermore, with (51) we estimate $\gamma_{res} \sim 10^3$ and the LMA MSW oscillation solution is clearly in the adiabatic regime. Other interesting features of the LMA MSW solution emerge by considering the limit of increasing neutrino energy:

$$E \uparrow \Rightarrow \begin{cases} x_{res} \text{ moves outward,} \\ \theta_m(x_0) \to \pi/2 \,, \\ P_{e1}^S \to 0, \; P_{e2}^S \to 1 \,, \\ P_{\nu_e \to \nu_e} \to \frac{1}{2}\left(1 - \cos 2\theta\right) = \sin^2\theta \,. \end{cases} \tag{53}$$

The first line follows from (25) and the fact that $N_e(x)$ is monotonically decreasing from the center to the edge of the sun.*** The other properties in (53) are read off from (38), (42) and (43), respectively. We remind the reader that the latter limit does not include earth matter effects, therefore, it holds only during the day, just as (43). Note that from $P_{e2}^S \to 1$ we conclude that for large enough solar neutrino energies we have a transition

$$|\nu_e\rangle_{\text{core}} \to |\nu_2\rangle_{\text{edge}} \,. \tag{54}$$

In Fig. 5 the solar ν_e survival probability is depicted. For historical reasons, $P_{\nu_e \to \nu_e}$ is depicted not only for the LMA MSW, but also the LOW and vacuum oscillation[†] probabilities which are both ruled out now. The dotted lines refer to daytime, i.e., when earth matter has no effect. The LMA MSW best fit of [49] is given by $\Delta m^2 = 5.0 \times 10^{-5}$ eV^2 and $\tan^2\theta = 0.42$ (after the SNO but before the KamLAND result). Let us now make a small numerical exercise. From the best fit we get $\sin^2\theta = 0.30$, which should give $P_{\nu_e \to \nu_e}$ for large E. Indeed, looking at the right end of the dotted curve in the right LMA panel where $E = 15$ MeV, we find excellent agreement. We see that neutrino energies around 10 MeV are already "large" in the sense of the limit (53). On the other hand, let us take the limit $E \to 0$. In that limit the neutrino does not go through a resonance because (25) cannot be fulfilled. Therefore, because of the short oscillation length (52), we expect an averaged survival probability $P_{\nu_e \to \nu_e} = 1 - \frac{1}{2}\sin^2 2\theta$ (see (9) and (11)).[‡] Numerically, we obtain $1 - \frac{1}{2}\sin^2 2\theta = 0.58$, in excellent agreement with the left ends of the LMA panels. We have thus demonstrated here that qualitatively the LMA MSW solution can be quite easily understood.

We have not discussed earth matter effects here. Looking at the right LMA panel in Fig. 5, we see that, at energies $E \gtrsim 10$ MeV, during the night when earth matter effects are operative the ν_e survival probability is a little larger than during the day. This introduces a small day–night asymmetry due

---

*** Note that a neutrino can go twice through a resonance if it is produced in that half of the sun which looks away from the earth.

† Here the oscillation length is of the order of the distance between sun and earth.

‡ We thank D.P. Roy for a discussion on this point.

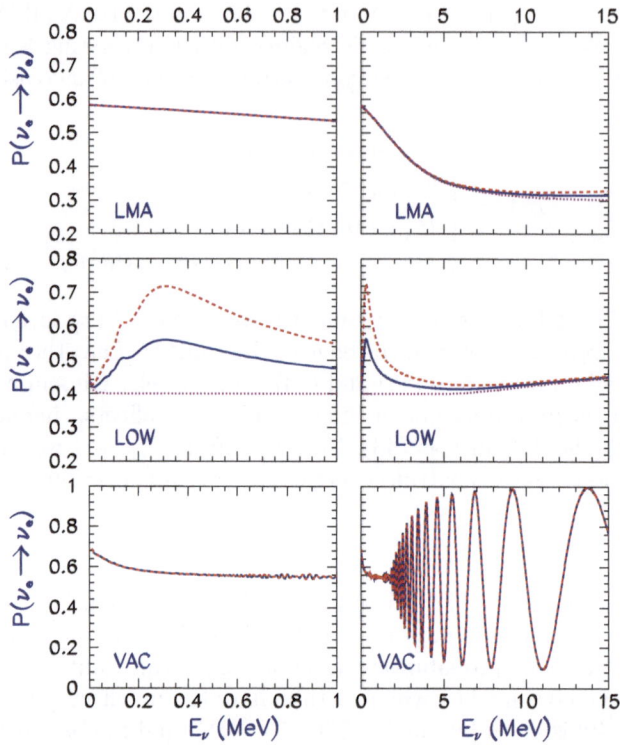

Fig. 5. Comparison of the survival probabilities for LMA MSW, LOW and vacuum solar neutrino oscillations for an electron neutrino created in the center of the sun. The figure has been taken from [49]. The dotted lines refer to the daytime, whereas the dashed line includes ν_e regeneration during the night, when earth matter is effective. The full line refers to the average survival probability.

to ν_e regeneration in earth matter. For a qualitative understanding of this effect see [50].

Small Δm^2 and the limit of strong non-adiabaticity: Here we want to discuss the question *How do solar neutrinos approach vacuum oscillations in the limit of small Δm^2 ?* This is a non-trivial question because for small Δm^2 there is always a point where the resonance condition (25) holds. Of course, this is a purely academic question since we know that solar neutrinos follow the LMA MSW solution for which the resonance does play a prominent role. It is, however, an interesting question from the point of view of neutrino evolution in matter in general.

When we speak of vacuum oscillations of solar neutrinos we mean mass-squared differences of the order of $\Delta m^2 \sim 10^{-10}$ eV^2 for which the oscillation length is of the order of the distance between sun and earth. From the resonance condition (25) we read off that in this case the resonance is very close to the solar edge (in this context, see $N_e(x)$ in [5]). From (38) it follows that $\theta_m(x_0) = \pi/2$ and that $\dot{\theta}_m = 0$ holds until very close to the resonance. It is easy to check that the width of the resonance, i.e. the distance where the change in $A(x)$ is of the order of $\Delta m^2 \cos 2\theta$, is roughly r_0, where the electron density $N(x)$ behaves like (50). Close to the edge of the sun the electron density drops steeper and the resonance width is smaller. Then one can check that while the neutrino crosses the resonance and also between the resonance and the solar edge one can approximate the phase φ of (45) by a constant, which can be taken as its value at the solar edge. This is just the opposite of adiabaticity – see the discussion after (44). For constant φ the solution of (44) is given by

$$\begin{pmatrix} a_1(x) \\ a_2(x) \end{pmatrix} = \begin{pmatrix} \cos \Delta\theta_m & -e^{i\varphi} \sin \Delta\theta_m \\ e^{-i\varphi} \sin \Delta\theta_m & \cos \Delta\theta_m \end{pmatrix} \begin{pmatrix} a_1(x_0) \\ a_2(x_0) \end{pmatrix} \qquad (55)$$

with $\Delta\theta_m \equiv \theta_m(x) - \theta_m(x_0)$. In the case under discussion, the solution (55) is valid from just before the resonance till the solar edge and in the beginning of the vacuum. The initial conditions are $\theta_m(x_0) = \pi/2$ and $a_1(x_0) = 0$, $a_2(x_0) = 1$. Then with (55), at the solar edge, we get $a_1 = e^{i\varphi} \cos \theta$, $a_2 = \sin \theta$. Plugging this result into the full solution (39), we find what we expect: After the resonance, i.e. from the the solar edge onward, neutrinos perform ordinary vacuum oscillations; even the matter effects accumulated in the phases $\varphi_{1,2}$ cancel. Inside the sun before the resonance, oscillations are completely suppressed because of the matter effects, due to $A \gg \Delta m^2 |\cos 2\theta|$. For further details see [48].

2.4   3-Neutrino Oscillations and the Mixing Matrix

Neutrino mass spectra: Up to now all possible neutrino mass spectra, i.e. hierarchical, inverted hierarchy, degenerate, etc., are compatible with present data. By convention we have stipulated $m_1 < m_2$ or $\Delta m^2_{21} \equiv \Delta m^2_\odot > 0$, where we have used the definition $\Delta m^2_{jk} = m^2_j - m^2_k$. Thus there are two possibilities for m_3 according to $\Delta m^2_{31} \gtrless 0$. Correspondingly, there are the two types of spectra depicted in Fig. 6: The "normal" and the "inverted" spectrum [51]. The hierarchical spectrum emerges in the limit $m_1 \to 0$ of the normal spectrum, whereas the "inverted" hierarchy is obtained from the inverted spectrum with $m_3 \to 0$.

In the 3-neutrino case the atmospheric mass-squared difference is not uniquely defined. If by convention we use for Δm^2_{atm} the largest possible mass-squared difference, then for the normal spectrum we obtain $\Delta m^2_{\text{atm}} = \Delta m^2_{31}$ and for the inverted case we have $\Delta m^2_{\text{atm}} = \Delta m^2_{23}$.

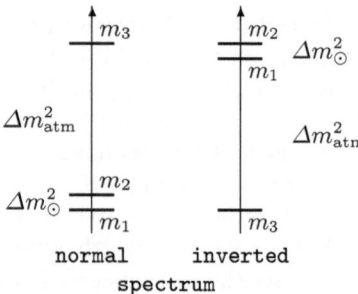

Fig. 6. The normal and the inverted 3-neutrino mass spectrum.

The mixing matrix: The most popular parameterization of the neutrino mixing matrix is given by [52]

$$U = U_{23}U_{13}U_{12} \tag{56}$$

with

$$U_{23} = \begin{pmatrix} 1 & 0 & 0 \\ 0 & c_{23} & s_{23} \\ 0 & -s_{23} & c_{23} \end{pmatrix}, \tag{57}$$

$$U_{13} = \begin{pmatrix} c_{13} & 0 & s_{13}e^{-i\delta} \\ 0 & 1 & 0 \\ -s_{13}e^{i\delta} & 0 & c_{13} \end{pmatrix}, \tag{58}$$

$$U_{12} = \begin{pmatrix} c_{12} & s_{12} & 0 \\ -s_{12} & c_{12} & 0 \\ 0 & 0 & 1 \end{pmatrix}. \tag{59}$$

The mixing angle $\theta_{12} \equiv \theta_{\odot}$ is probed in the solar neutrino experiments and in the KamLAND experiment. The mixing angle $\theta_{23} \equiv \theta_{\mathrm{atm}}$ is the relevant angle in atmospheric and LBL neutrino oscillation experiments. Up to now now effects of a non-zero angle θ_{13} have not been seen.

3-neutrino oscillations: The principle of 3-neutrino oscillations is a consequence of $\Delta m_{\odot}^2 \ll \Delta m_{\mathrm{atm}}^2$ and can be summarized in the following way.

$\Delta m_{\mathrm{atm}}^2 L/(2E) \sim 1 \Rightarrow \Delta m_{\odot}^2 L/(2E) \ll 1$ Solar ν oscillations frozen in atm./LBL ν osc.!

$\Delta m_{\odot}^2 L/(2E) \sim 1 \Rightarrow \Delta m_{\mathrm{atm}}^2 L/(2E) \gg 1$ Atm. ν oscillations averaged in solar ν osc.!

Since, at present, $\theta_{13} = 0$ is compatible with all available neutrino data, only an upper bound on this angle can be extracted. The two most important results in this context are:

- The non-observation of ν_e disappearance in the CHOOZ [20] and Palo Verde [21] experiments;
- The non-observation of atmospheric ν_e or $\bar{\nu}_e$ disappearance [4].

Also solar neutrino data have an effect on θ_{13}, although small, because in the 3-neutrino case the solar survival probability is given by

$$P^{\odot}_{\nu_e \to \nu_e} \simeq c^4_{13} \, P^{(2)}_{\nu_e \to \nu_e}(\Delta m^2_{21}, \theta_{12}, c^2_{13}N_e) + s^4_{13} \,, \tag{60}$$

where $P^{(2)}_{\nu_e \to \nu_e}$ is a 2-neutrino probability, calculated with the electron density $c^2_{13}N_e$ instead of N_e. The result of fits to the neutrino data is [53]

$$\sin^2 \theta_{13} < 0.05 \text{ at } 3\sigma \,. \tag{61}$$

As evident from the parameterization (56), CP violation disappears from neutrino oscillations for $\theta_{13} \to 0$.

In atmospheric neutrino experiments the mixing parameter which is extracted is

$$\sin^2 2\theta_{23} \simeq 4|U_{\mu 3}|^2(1 - |U_{\mu 3}|^2) \,, \tag{62}$$

where we have taken into account the smallness of θ_{13}. The Super-Kamiokande experiment finds a result [54] compatible with maximal mixing:

$$\sin^2 2\theta_{23} > 0.92 \text{ at } 90\% \text{ CL.} \tag{63}$$

This leads to $0.60 < |U_{\mu 3}| < 0.80$. Note that $\theta_{23} = 45°$ corresponds to $|U_{\mu 3}| = 1/\sqrt{2} \simeq 0.707$.

Now we want to discuss the states into which solar and atmospheric neutrinos are transformed. In the limit $\theta_{13} \to 0$ or $U_{13} \to 1$ this discussion is most transparent. In that limit, in terms of neutrino states, atmospheric neutrino oscillations are given by $|\nu_\mu\rangle \to |\nu_\tau\rangle$, $|\bar{\nu}_\mu\rangle \to |\bar{\nu}_\tau\rangle$, while ν_e and $\bar{\nu}_e$ do not oscillate because their oscillation amplitude is given by $\sin^2 2\theta_{13}$. In the context of (35) the question was raised to which neutrino state the amplitude a_x belongs. For the evolution of neutrino states in the sun, the mass eigenstate $|\nu_3\rangle = s_{23}|\nu_\mu\rangle + c_{23}|\nu_\tau\rangle$ is an approximate eigenstate of H^{mat}_ν because $|\Delta m^2_{31}| \gg A$ (see (24)). Therefore, the initial state $|\nu_e\rangle_\odot$ must transform with probability $P^{\odot}_{\nu_e \to \nu_e}$ into the state orthogonal to $|\nu_3\rangle$, from where it follows that (without earth regeneration effects)

$$|\nu_e\rangle_\odot \to -c_{23}|\nu_\mu\rangle + s_{23}|\nu_\tau\rangle \,. \tag{64}$$

Note that Equation (60) has been derived by using the approximate eigenvector property of $|\nu_3\rangle$, but corrections for non-zero θ_{13} have been made.

The primary goal of LBL neutrino oscillation experiments is the check of atmospheric neutrino results, observation of ν_τ appearance and the oscillatory behaviour in L/E [4]. The demonstration of the latter property would give us the final proof that atmospheric ν_μ and $\bar{\nu}_\mu$ disappearance is really an effect of oscillations. Let us make a list what we would wish to gain from LBL (K2K with $L = 250$ km, MINOS and CERN-Gran Sasso with $L = 730$ km) and very LBL experiments:

* Precision measurements of Δm^2_{atm} and θ_{23};
* Measurement of θ_{13};
* Verification of matter effects and distinction between normal and inverted mass spectra;
* Effects of $\Delta m^2_\odot$ and CP violation.

We want to stress the difficulty of measuring CP violation in neutrino oscillations: Such an effect is only present if $\sin\theta_{13} \neq 0$ and if the experiment is sensitive to both mass-squared differences, atmospheric and solar, thus the measurements must be accurate to such a degree that they invalidate the principle of 3-neutrino oscillations mentioned in the beginning of this subsection. Furthermore, the earth matter background in (very) LBL experiments fakes effects of CP violation which have to be separated from true CP violation.

Very LBL experiments are planned with super neutrino beams [55] and neutrino factories [56]. Super neutrino beams are conventional but very-high-intensity and low-energy ($E \sim 1$ GeV) beams, with the neutrino detector slightly off the beam axis; the JHF–Kamioka Neutrino Project with $L = 295$ km is scheduled to start in 2007 [55]. Neutrino factories are muon storage rings with straight sections from where well-defined neutrino beams from muon decay emerge. According to the list above, measuring $\nu_e \to \nu_\mu$ and $\bar{\nu}_e \to \bar{\nu}_\mu$ transitions is of particular interest; note that θ_{13} is the only angle in the neutrino mixing matrix which has not been measured up to now. One can easily convince oneself that this task is tackled by measuring the so-called "wrong-sign" muons; e.g., $\mu^-_{\text{storage}} \to \mu^+_{\text{detector}}$ corresponds to $\bar{\nu}_e \to \bar{\nu}_\mu$. Neutrino factories are under investigation. Optimization conditions for running such a machine are considered in the ranges $20 \lesssim E_\mu \lesssim 50$ GeV and $L \gtrsim 3000$ km. For further references see, e.g., [57–60] and citations therein.

3   Dirac Versus Majorana Neutrinos

In this section we focus, in particular, on Majorana neutrinos. Apart from the vanishing electric charge, also most of the popular extensions of the SM suggest that neutrinos have Majorana nature. Majorana particles have several interesting features, which make them quite different from Dirac particles. Unfortunately, in practice, due to the smallness of the masses of the light neutrinos, it is quite difficult to distinguish between both natures: Neutrinoless $\beta\beta$-decay seems to be the only promising road so far. In any case, neutrino oscillations do not distinguish between Dirac and Majorana neutrinos, since there only the states with negative helicity enter, where this distinction is irrelevant. Though family lepton numbers must be violated for transitions $\nu_\alpha \to \nu_\beta$, the total lepton number remains conserved.

3.1 Free Fields

With two independent chiral 4-spinor fields $\psi_{L,R}$ one can construct the usual *Lorentz-invariant* bilinear for Dirac fields which is called mass term:

$$\text{Dirac:} \quad -m\left(\bar{\psi}_R\psi_L + \text{H.c.}\right) = -m\bar{\psi}\psi \quad \text{with} \quad \psi = \psi_L + \psi_R. \tag{65}$$

With only one chiral 4-spinor ψ_L one obtains nevertheless a *Lorentz-invariant* bilinear with the help of the charge-conjugation matrix C:

$$\text{Majorana:} \quad \frac{1}{2}m\psi_L^T C^{-1}\psi_L + \text{H.c.} = -\frac{1}{2}m\bar{\psi}\psi \text{ with } \psi = \psi_L + (\psi_L)^c. \tag{66}$$

Note that from ψ_L one obtains a right-handed field with the charge-conjugation operation $(\psi_L)^c \equiv C\gamma_0^T\psi_L^*$; in contrast to the Dirac case, this right-handed field is *not* independent of the left-handed field.

The equation for free Dirac or Majorana fields in terms of the above defined spinors ψ is the same:

$$\text{Dirac and Majorana:} \quad (i\gamma^\rho\partial_\rho - m)\psi = 0. \tag{67}$$

In the formalism used here the Majorana nature is hidden in the

$$\text{Majorana condition:} \quad \psi = \psi^c. \tag{68}$$

Thus the solution of (67) is found in the same way for both fermion types. However, for Majorana neutrinos the condition (68) is imposed on the solution. This leads to the following observations.

Dirac fermions: The field ψ contains annihilation and creation operators a, $b^\dagger$, respectively, with independent operators a, b, and thus a Dirac fermion field has particles and antiparticles with positive and negative helicities.

Majorana fermions: Because of the condition (68) the annihilation and creation operators are a, $a^\dagger$, respectively, and we have only particles with positive and negative helicities.

The mass terms (65) and (66) are written in the way as they appear in the Lagrangian. The field equation (67) is obtained from the Lagrangian by variation with respect to the independent fields. In this procedure the factor $1/2$ in the Majorana mass term is canceled by the factor of 2 which occurs in the variation of the mass term because the fields to the left and to the right of C^{-1} are identical (see the first term in (66)).

Up to now we have discussed the case of one neutrino with mass $m > 0$. Let us now consider n neutrinos. Then in the Dirac case we have the following mass term:

$$\text{Dirac:} \quad -(\bar{\nu}_R \mathcal{M}\nu_L + \text{H.c.}) = -\bar{\nu}'\hat{m}\nu'. \tag{69}$$

Here, $\mathcal{M}$ is an *arbitrary* complex $n \times n$ matrix and the fields $\nu_{L,R}$ are vectors containing n 4-spinors. In order to arrive at the diagonal and positive matrix $\hat{m}$ we use the following theorem concerning bidiagonalization in linear algebra.

Theorem 1. *If $\mathcal{M}$ is an arbitrary complex $n \times n$ matrix, then there exist unitary matrices $U_{L,R}$ with $U_R^\dagger \mathcal{M} U_L = \hat{m}$ diagonal and positive.*

Applying this theorem we obtain the

$$\text{physical Dirac fields} \quad \nu' = \nu'_L + \nu'_R \quad \text{with} \quad \nu_{L,R} = U_{L,R}\nu'_{L,R}. \quad (70)$$

Note that the $U(1)$ invariance of the mass term under $\nu_{L,R} \to e^{i\alpha}\nu_{L,R}$ corresponds to total lepton number conservation, provided the rest of the Lagrangian respects this symmetry as well.

Switching to the Majorana case we have the following mass term:

$$\text{Majorana:} \quad \frac{1}{2}\nu_L^T C^{-1}\mathcal{M}\nu_L + \text{H.c.} = \frac{1}{2}\nu_L'^T C^{-1}\hat{m}\,\nu'_L + \text{H.c.} =$$
$$-\frac{1}{2}\bar{\nu}'\hat{m}\nu'. \quad (71)$$

Now the mass matrix $\mathcal{M}$ is a complex matrix which fulfills

$$\mathcal{M}^T = \mathcal{M}. \quad (72)$$

This follows from the anticommutation property of the fermionic fields and $C^T = -C$. The diagonalization of the mass term proceeds now via a theorem of I. Schur [61].

Theorem 2. *If $\mathcal{M}$ is a complex symmetric $n \times n$ matrix, then there exists a unitary matrix U_L with $U_L^T \mathcal{M} U_L = \hat{m}$ diagonal and positive.*

With this theorem we obtain the

$$\text{physical Majorana fields} \quad \nu' = \nu'_L + (\nu'_L)^c \quad \text{with} \quad \nu_L = U_L\nu'_L. \quad (73)$$

The mass term (71) not only violates individual lepton family numbers just as the Dirac mass term (69), but it also violates the total lepton number $L = \sum_\alpha L_\alpha$.

3.2   Majorana Neutrinos and CP

It is easy to check that the Majorana mass term corresponding to the mass eigenfields, the second expression in (71) (now we drop the primes on the fields ν_L), is invariant under [11,62]

$$\nu_{jL} \to \rho_j\, i\, C\nu_{jL}^* \quad \text{with} \quad \rho_j^2 = 1. \quad (74)$$

Thus Majorana neutrinos have imaginary CP parities. One can check that $\nu_j = \nu_{jL} + (\nu_{jL})^c$ transforms into $\rho_j\, i\, C\nu_j^*$ under the transformation (74). If we supplement this CP transformation by

$$W_\mu^+ \to -\epsilon(\mu)W_\mu^- \quad \text{with} \quad \epsilon(\mu) = (-1)^{\delta_{0\mu}+1}, \quad (75)$$
$$\ell_{\alpha L} \to -C\ell_{\alpha L}^* \quad (\alpha = e, \mu, \tau) \quad (76)$$

and require that the CC interactions are CP-invariant then we are lead to

$$U_M = i\, U_M^* \hat{\rho} \quad \text{with} \quad \hat{\rho} = \text{diag}\,(\rho_1, \rho_2, \rho_3)\,. \tag{77}$$

We denote the mixing matrix for Majorana neutrinos by U_M. From (77) we derive

$$\text{CP invariance} \;\Rightarrow\; U_M = O\, e^{i\frac{\pi}{4}\hat{\rho}}\,, \tag{78}$$

where O is a real orthogonal matrix. In the CP-conserving case the matrix O is relevant for neutrino oscillations since the phases in U_M of (78) drop out due to the rephasing invariance of the oscillation probability (see Sect. 2.1). Note that one can extract an overall factor $\exp(i\pi/4)$ or $\exp(-i\pi/4)$ from the phase matrix in U_M and absorb it into the charged lepton fields. In this way, one obtains another commonly used form of U_M in the case of CP conservation.

For CP non-invariance one obtains

$$U_M = U e^{i\hat{\alpha}}\,. \tag{79}$$

The *Majorana phases* α_j cannot be removed by absorption into ν_{jL} because the Majorana mass term is not invariant under such a rephasing. However, one of the three phases can be absorbed into a the charged lepton fields, thus only two of the Majorana phases are physical. In neutrino oscillations only U (56) with the KM phase δ is relevant. The two Majorana phases play an important role in the discussion of neutrinoless $\beta\beta$-decay and can be related to the phases which appear in leptogenesis and the baryon asymmetry from leptogenesis [63] (for reviews see, for instance, [64]), though in general the low-energy CP phases are independent of the phases relevant in leptogenesis [65,66] (see also [67] and citations therein).

3.3   Neutrinoless $\beta\beta$-Decay

Up to now, the lepton number-conserving $\beta\beta$-decay $(Z, A) \rightarrow (Z + 2, A) + 2e^- + 2\bar{\nu}_e$ has been observed in direct experiments with seven nuclides [68]. There is also an intensive search for the neutrinoless $\beta\beta$ or $(\beta\beta)_{0\nu}$ decay $(Z, A) \rightarrow (Z + 2, A) + 2e^-$, where the total lepton number $L = L_e + L_\mu + L_\tau$ is violated ($\Delta L = 2$). The distinct signal for such a decay is given by the two electrons each with energy $E_e = (M(Z, A) - M(Z + 2, A))/2$, i.e. half of the mass difference between mother and daughter nuclide. In contrast to $\beta\beta$-decay with neutrinos, there is no unequivocal experimental demonstration for such a decay. The most stringent limit comes from ^{76}Ge with $T_{1/2}^{0\nu} > 1.9 \times 10^{25}$ yr (90% CL) [69]. For reviews on $(\beta\beta)_{0\nu}$ decay experiments see [68,70].

There are several mechanisms on the quark level which induce two transitions $n \rightarrow p$ in a nucleus without emission of neutrinos. Here we discuss some of the popular ones (see [71] for a review of $(\beta\beta)_{0\nu}$ decay mechanisms).

- **Effective Majorana neutrino mass**: This mechanism uses the Majorana nature of neutrinos and can schematically be depicted in the following way:

$$\left.\begin{array}{l} d \to u + e^- + \bar{\nu}_e \\ d \to u + e^- + \bar{\nu}_e \end{array}\right] \quad \text{Wick contraction}$$

This Wick contraction can be performed because $\nu_j^c = \nu_j$ and, therefore, the electron-neutrino field can be written as

$$\nu_{eL} = \sum_j (U_M)_{ej} \nu_{jL} = \sum_j (U_M)_{ej} \frac{1 - \gamma_5}{2} C(\bar{\nu}_j)^T . \tag{80}$$

Thus one actually contracts ν_j with $\bar{\nu}_j$ $\forall j$. In other words, the relevant neutrino propagator is given by the expression

$$\langle 0|T\nu_{eL}(x_1)\nu_{eL}^T(x_2)|0\rangle = \\ -\sum_j (U_M)_{ej}^2 m_j\, i \int \frac{d^4 p}{(2\pi)^2} \frac{e^{-ip \cdot (x_1 - x_2)}}{p^2 - m_j^2} \frac{1 - \gamma_5}{2} C . \tag{81}$$

One can show that it is allowed to neglect m_j^2 in the denominator of the ν_j propagators [70], which leads then to the *effective Majorana neutrino mass*

$$\langle m \rangle \equiv \sum_j (U_M)_{ej}^2 m_j . \tag{82}$$

For this mechanism of $(\beta\beta)_{0\nu}$ decay, the decay amplitude is proportional to $|\langle m \rangle|$.

The present lower bounds on $T_{1/2}^{0\nu}$ for ^{76}Ge lead to an upper on $|\langle m \rangle|$ varying between 0.33 and 1.35 eV, depending on the models used to calculate the nuclear matrix element [69,72]. Future experiments want to probe $|\langle m \rangle|$ down to a few $\times$ 0.01 eV [70].

In the case of CP conservation (see (78)) the effective Majorana neutrino mass is given by $\langle m \rangle = i \sum_j O_{ej}^2 \rho_j m_j$. For opposite signs of $\rho_j = \pm 1$, i.e. opposite CP parities, cancellations in $\langle m \rangle$ naturally occur.

- **Intermediate doubly charged scalar**: In left-right symmetric models, doubly charged scalars occur in the scalar gauge triplets and induce $(\beta\beta)_{0\nu}$ decay via the mechanism [71] depicted symbolically in the following way:

$$\left.\begin{array}{l} d \to u + S^- \searrow \\ d \to u + S^- \nearrow \end{array}\right\rangle H^{--} \to e^- e^-$$

- **SUSY with R-parity violation**: Within this framework, several mechanisms for inducing $(\beta\beta)_{0\nu}$ decay exist [71]. The gluino $(\tilde{g})$ exchange mechanism uses the Majorana nature of gauginos [73]:

$$d \to e^- + \tilde{u}$$
$$\searrow u + \tilde{g}$$
$$\left.\begin{array}{c} \\ \end{array}\right] \text{Wick contraction}$$
$$\nearrow u + \tilde{g}$$
$$d \to e^- + \tilde{u}$$

In this picture, $\tilde{u}$ is an up squark.

For the check of the Majorana nature of neutrinos, $(\beta\beta)_{0\nu}$ decay is most realistic possibility. Another line which is pursued is the search for $\bar{\nu}_e$ from the sun [74]. Both types of experiments aim at finding $|\Delta L| = 2$ processes.

In view of several mechanisms for $(\beta\beta)_{0\nu}$ decay where some even do not require neutrinos, the question arises if an experimental confirmation of this decay really signals Majorana nature of neutrinos. The question was answered affirmatively by Schechter and Valle [75], and Takasugi [76]. Their statement is the following: *If neutrinoless $\beta\beta$-decay exists, then neutrinos have Majorana nature, irrespective of the mechanism for $\beta\beta$-decay.* We follow the argument of Takasugi and show the following: *If $(\beta\beta)_{0\nu}$ decay exists, then a Majorana neutrino mass term cannot be forbidden by a symmetry.*
Proof: We assume for simplicity that there is no neutrino mixing. As will be seen the arguments can easily be extended to the mixing case. We introduce phase factors η and assume the symmetry

$$u_L \to \eta_u u_L \,, \quad d_L \to \eta_d d_L \,, \quad W^+ \to \eta_W W^+ \,. \tag{83}$$
$$e_L \to \eta_e e_L \,, \quad \nu_{eL} \to \eta_\nu \nu_{eL} \,,$$

Invariance of the weak interactions under this symmetry give the following relations:

$$\bar{\nu}_{eL} \gamma^\rho e_L W_\rho^+ \Rightarrow \eta_\nu^* \eta_e \eta_W = 1 \,, \tag{84}$$

$$\bar{u}_L \gamma^\rho d_L W_\rho^+ \Rightarrow \eta_u^* \eta_d \eta_W = 1 \,. \tag{85}$$

Now we assume the existence of $(\beta\beta)_{0\nu}$ decay, which introduces a further relation:

$$d_L + d_L \to u_L + u_L + e_L + e_L \Rightarrow (\eta_d^* \eta_u \eta_e)^2 = 1 \tag{86}$$

Equations (84) and (85) together lead to $\eta_\nu = \eta_d^* \eta_u \eta_e$. Thus with (86) we arrive at $\eta_\nu^2 = 1$, and the mass term $\propto \nu_{eL}^T C^{-1} \nu_{eL}$ cannot be forbidden by $\nu_{eL} \to \eta_\nu \nu_{eL}$. Q.E.D.

The above argument does not necessarily show that a Majorana neutrino mass term must appear, but in general this will be the case for renormalizeability reasons: All terms of dimension 4 or less which are compatible with the symmetries of the theory have to be included in the Lagrangian.

3.4   Absolute Neutrino Masses

With neutrino oscillations, the lightest neutrino mass which we denote by m_0 cannot be determined. Nowadays three sources of information on m_0 are used: ^{3}H decay, $(\beta\beta)_{0\nu}$ decay and cosmology [77]. For an extensive review see [51].

Tritium and neutrinoless $\beta\beta$-decays: In the decay ^{3}H $\rightarrow$ ^{3}He $+ e^- + \bar{\nu}_e$, with an energy release $E_0 \simeq 18.6$ keV, the region near the endpoint of the electron recoil energy spectrum is investigated. Deviations from the ordinary β spectrum with a massless $\bar{\nu}_e$ are searched for. Assuming that the $\bar{\nu}_e$ has a mass m_β, the Mainz and Troitsk experiments have both obtained an upper limit $m_\beta < 2.2$ eV at 95% CL. The future KATRIN experiment plans to have a sensitivity $m_\beta \gtrsim 0.35$ eV [4].

Since we know that the electron neutrino field is a linear combination of neutrino mass eigenfields because of neutrino mixing (see (2)), the ^{3}H decay rate is actually a sum over the decay rates of the massive neutrinos multiplied by $|U_{ej}|^2$. Thus the question arises for meaning of m_β which is extracted from the data. Farzan and Smirnov [78] have argued in the following way. Firstly, the bulk of data from which the bound on m_β is derived does *not* come from the very end of recoil spectrum. Secondly, the energy resolution in ^{3}H experiments, even for the KATRIN experiment, is significantly coarser than $\sqrt{\Delta m_{\mathrm{atm}}^2}$. Therefore, the three masses m_j can effectively be replaced by one mass m_β and the different expressions for m_β found in the literature, e.g. $m_\beta \equiv \sum_j |U_{ej}|^2 m_j$, are all equivalent. The simplest choice is $m_\beta = m_0$ (for other papers on this subject see citations in [78]). Thus the results of the ^{3}H decay experiments give information on the lightest neutrino mass m_0.

In order to make contact between m_0 and $(\beta\beta)_{0\nu}$ decay, the following assumptions are made: Clearly, one has to assume Majorana neutrino nature, but also that the dominant mechanism for $(\beta\beta)_{0\nu}$ decay proceeds via the effective Majorana neutrino mass $|\langle m \rangle|$ (82), which is then expressed as a function of m_0 and the parameters of the Majorana neutrino mixing matrix U_M (79). Note that this mechanism *must* be present for Majorana neutrinos, it might only be that it is not the dominant one. For the **normal** spectrum we have $m_0 = m_1$ and $m_0 = m_3$ for the **inverted** spectrum (see Fig. 6).

Normal spectrum:

$$|\langle m \rangle| = \left| \left(m_0\, c_{12}^2 + \sqrt{m_0^2 + \Delta m_\odot^2}\, s_{12}^2\, e^{i\beta_1} \right) c_{13}^2 + \right.$$

$$\left. \sqrt{m_0^2 + \Delta m_{\mathrm{atm}}^2}\, s_{13}^2\, e^{i\beta_2} \right| ; \tag{87}$$

Inverted spectrum:

$$|\langle m \rangle| = \left| \left(\sqrt{m_0^2 + \Delta m_{\mathrm{atm}}^2 - \Delta m_\odot^2} \; c_{12}^2 \, e^{i\beta_1} + \right. \right.$$

$$\left. \left. \sqrt{m_0^2 + \Delta m_{\mathrm{atm}}^2} \; s_{12}^2 \, e^{i\beta_2} \right) c_{13}^2 + m_0 \, s_{13}^2 \right|. \quad (88)$$

In these expressions, β_1 and β_2 are general CP-violating phases which arise according to the form of U_M.[§] For numerical estimates it is useful to have in mind that $\sqrt{\Delta m_{\mathrm{atm}}^2} \sim 0.05$ eV and $\sqrt{\Delta m_\odot^2} \sim 0.008$ eV. For references see, for instance, [79,80] and citations therein.

In order to plot $|\langle m \rangle|$ against m_0, the values of neutrino oscillation parameters are used as input. The phases $\beta_{1,2}$ are free parameters; for CP conservation the possible values of $\beta_{1,2}$ are zero or π. The plot of Fig. 7 has been taken from [79]. In the upper panel, best fit values of the neutrino oscillation parameters are used, whereas in the lower two panels the uncertainty in these parameters is taken into account. The lower band on the left sides of these panels correspond to the normal spectrum, whereas the upper band to the inverted spectrum.

What does this figure teach us? One can use experimental upper bounds on $|\langle m \rangle|$ and m_0 and compare with the allowed regions in the plot. It has been put forward in [81] that such a comparison, even if $(\beta\beta)_{0\nu}$ decay is not found experimentally, might reveal the neutrino nature: If the minimum of $|\langle m \rangle|$ predicted from m_β and the oscillation parameters exceeds the experimental upper bound on $|\langle m \rangle|$, then the neutrino must have Dirac nature. There are, however, several difficulties with such a procedure: Other mechanisms for $(\beta\beta)_{0\nu}$ decay could destroy the validity of such a comparison; m_β will not reach the non-degenerate neutrino mass region even with the KATRIN experiment, and the same difficulty applies to the region with $|\langle m \rangle|$ below 0.01 eV; moreover, imprecise determination of the oscillation parameters blur the picture as seen from the lower two panels of Fig. 7. However, the chances for this approach would be good, if the KATRIN experiment would find a non-zero m_β. On the other hand, assuming the validity of the Majorana hypothesis, Fig. 7 might give us information about the type of neutrino mass spectrum if $|\langle m \rangle|$ is measured or a sufficiently stringent experimental bound on it is derived [79].

Neutrino masses and cosmology: A very interesting bound on the sum over all neutrino masses is provided by the large-scale structure of the universe [82] and the temperature fluctuations of the cosmic microwave background (CMB) [83]. Usually, energy densities ρ_i in the universe are given as fractions $\Omega_i \equiv \rho_i/\rho_{\mathrm{cr}}$ of the critical energy density ρ_{cr} of the universe. Today's energy

---

[§] They are simple functions of the Majorana phases and of δ.

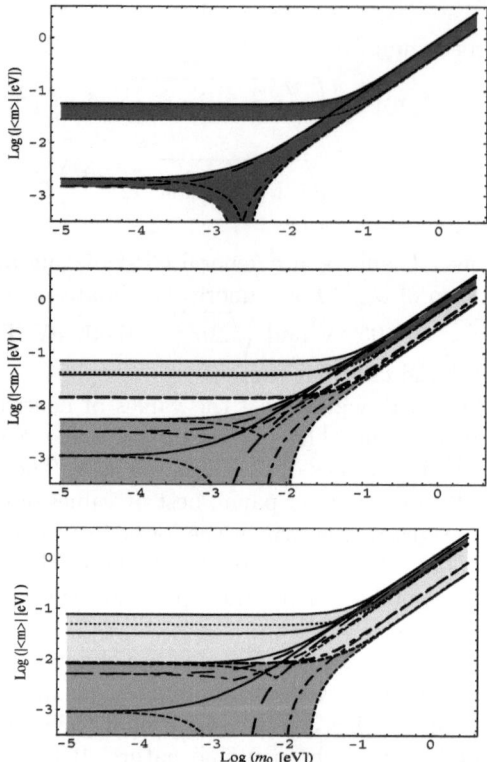

Fig. 7. Effective Majorana neutrino mass $|\langle m\rangle|$ versus lightest mass m_0. The figure is taken from [79]. In the upper panel best fit values for the neutrino oscillation parameters have been used and only the phases β_1 and β_2 have been varied. (A non-zero best fit value for s_{13} is used in this plot.) In the lower two panels the uncertainty in the determination of the oscillation parameters has been taken into account. For the details and for the precise meaning of the lines in the shaded regions, which correspond to CP-conserving choices of the phases $\beta_{1,2}$, see [79].

density of non-relativistic neutrinos and antineutrinos is given by (see, for instance, [84])

$$\Omega_\nu h^2 = \sum_j m_j/(93.5\,\mathrm{eV})\,, \tag{89}$$

where $h \simeq 0.7$ is the Hubble constant in units of $100\ \mathrm{km\,s^{-1}\,Mpc^{-1}}$. Of the three active neutrinos, at least two of them are non-relativistic today since $\sqrt{\Delta m_\odot^2} \sim 0.008$ eV and $\sqrt{\Delta m_{\mathrm{atm}}^2} \gg \sqrt{\Delta m_\odot^2}$, while the neutrino temperature today corresponds to 1.7×10^{-4} eV [84].

Roughly speaking, an upper bound on Ω_ν is obtained in the following way. Hot dark matter tends to erase primordial density fluctuations. The theory of structure formation together with data on large-scale structures (distribution of galaxies and galaxy clusters), mainly from the 2 degree field Galaxy Redshift Survey [82], gives an upper bound on Ω_ν/Ω_m, where Ω_m is the total matter density. On the other hand, data on the temperature fluctuations of the CMB can be evaluated with the so-called ΛCDM model (a flat universe with cold dark matter and a cosmological constant) and allows in this way to determine a host of quantities like h, Ω_m, Ω_b (the baryon density), etc. Since the quantities extracted in this way agree rather well with determinations based on different methods and assumptions [83], the ΛCDM model is emerging as the standard model of cosmology [85]. Evaluation of the recent CMB results of the Wilkinson Microwave Anisotropy Probe (WMAP) and of the large-scale structure data give the impressive bound [83] $\Omega_\nu h^2 <$ 0.0076 (95% CL); with (89) this bound is reformulated in terms of neutrino masses:

$$\sum_j m_j < 0.7 \text{ eV } (95\%\text{CL}). \tag{90}$$

For a single neutrino we have thus $m_j < 0.23$ eV, since neutrinos with masses in the range of a few tenths eV have to be degenerate.

Note that the ΛCDM model gives the result $\Omega_m \sim 0.3$ and $\Omega_\Lambda \sim 0.7$; the latter quantity is the energy density associated with the cosmological constant Λ. Furthermore, one finds $\Omega_\nu \lesssim 0.015$ and $\Omega_b \sim 0.04 \div 0.05$, thus the main contribution to Ω_m must consist of hitherto undetermined dark matter components.

3.5 Neutrino Electromagnetic Moments

The effective Hamiltonians for neutrinos with magnetic moments and electric dipole moments (electromagnetic moments) are given by the following expressions:

$$\text{Dirac: } \mathcal{H}_{\text{em}}^D = \frac{1}{2}\bar{\nu}_R \lambda \sigma^{\mu\nu} \nu_L F_{\mu\nu} + \text{H.c.,} \tag{91}$$

$$\text{Majorana: } \mathcal{H}_{\text{em}}^M = -\frac{1}{4}\nu_L^T C^{-1} \lambda \sigma^{\mu\nu} \nu_L F_{\mu\nu} + \text{H.c.} \tag{92}$$

In the Majorana case, from $(C^{-1}\sigma^{\mu\nu})^T = C^{-1}\sigma^{\mu\nu}$ and the anticommutation property of ν_L, it follows that $\lambda^T = -\lambda$ and only transition moments are allowed. In the Dirac case, the electromagnetic moment matrix λ is an arbitrary 3×3 matrix. Usually, a decomposition of λ is made into $\lambda = \mu - id$ with $\mu^\dagger = \mu$ being the MM matrix and $d^\dagger = d$ the EDM matrix. However, since the neutrino mass eigenstates are experimentally not accessible, the distinction MM/EDM is completely unphysical in terrestrial experiments. This holds also largely for solar neutrinos, though in the evolution of the solar neutrino state the squares of the neutrino masses enter, and one can shift phases from the mixing matrix to the electromagnetic moment matrix and vice versa, which makes the distinction MM/EDM phase-convention-dependent. For a thorough discussion of this point see [86].

Bounds on λ from solar and reactor neutrinos are obtained via elastic νe^- scattering where, due to the helicity flip in the Hamiltonians (91) and (92), the cross section is the sum of the weak and electromagnetic cross sections [87]:

$$\frac{d\sigma}{dT} = \frac{d\sigma_w}{dT} + \frac{d\sigma_{\text{em}}}{dT} \quad \text{with} \quad \frac{d\sigma_{\text{em}}}{dT} = \frac{\alpha^2 \pi}{m_e^2 \mu_B^2} \left(\frac{1}{T} - \frac{1}{E} \right) \mu_{\text{eff}}^2 . \tag{93}$$

In the most general form, the effective MM is given by [86]

$$\mu_{\text{eff}}^2 = a_-^\dagger \lambda^\dagger \lambda a_- + a_+^\dagger \lambda \lambda^\dagger a_+ . \tag{94}$$

In (93), T is the kinetic energy of the recoil electron, μ_B is the Bohr magneton, and the flavour 3-vectors $a_\pm$ describe the neutrino helicity states at the detector. The expression (94) is basis-independent, thus it does not matter in which basis, flavour or mass basis, the quantities λ, $a_\pm$ and the Hamiltonians (91) and (92) are perceived [86].

In the following we will consider reactor and solar neutrinos. If the detector is close to the reactor, one simply has (in the flavour basis) $a_- = 0$, $a_+ = (1, 0, 0)^T$, as the reactor emits antineutrinos with positive helicity. For solar neutrinos, the effective MM will, in general, depend on the neutrino energy E. Bounds on the neutrino electromagnetic moments have been obtained from both reactor [88] and solar neutrinos [89,90].

From now on, we concentrate on Majorana neutrinos, where the antisymmetric matrix λ contains only three complex parameters (in the Dirac case there are nine parameters). Thus, λ can be written as

$$\lambda_{\alpha\beta} = \varepsilon_{\alpha\beta\gamma} \Lambda_\gamma , \tag{95}$$

in either basis, flavour or mass basis. For the approximations used in the calculation of the effective MM for solar neutrinos and the LMA MSW solution, μ_{LMA}^2, we refer the reader to [91]. The result is given by

$$\mu_{\text{LMA}}^2(E) = |\mathbf{\Lambda}|^2 - |\Lambda_2|^2 + P_{e1}^S \left(|\Lambda_2|^2 - |\Lambda_1|^2 \right) . \tag{96}$$

The probability P_{e1}^S is defined before (32). For reactor neutrinos the effective MM is immediately obtained as

$$\begin{aligned}\mu_{\text{reactor}}^2 &= |\lambda_{e\mu}|^2 + |\lambda_{e\tau}|^2 = |\Lambda_\mu|^2 + |\Lambda_\tau|^2 \\ &= |\mathbf{\Lambda}|^2 - c_{12}^2 |\Lambda_1|^2 - s_{12}^2 |\Lambda_2|^2 - 2 s_{12} c_{12} |\Lambda_1||\Lambda_2| \cos \delta' , \end{aligned} \tag{97}$$

where the second line is derived from the first line by using $\mathbf{\Lambda} = U_M \tilde{\mathbf{\Lambda}}$. The phase δ' is composed of $\delta' = \arg(\Lambda_1^* \Lambda_2) + \alpha_2 - \alpha_1$ with the Majorana phases α_j defined in (79). Equation (95) defines the vectors $\mathbf{\Lambda} = (\Lambda_\alpha)$ and $\tilde{\mathbf{\Lambda}} = (\Lambda_j)$ in the flavour and mass basis, respectively. The length of $\mathbf{\Lambda}$ occurs in both effective moments, (96) and (97). Note, however, that

$$|\mathbf{\Lambda}|^2 = \frac{1}{2} \text{Tr} \left(\lambda^\dagger \lambda \right) \quad \Rightarrow \quad |\mathbf{\Lambda}| = |\tilde{\mathbf{\Lambda}}| , \tag{98}$$

i.e., the length of the vector of the electromagnetic moments is basis-independent.

In [91] a bound on $|\mathbf{\Lambda}|$ is derived, i.e., *all* three transition moments are bounded by using as input the rates of the solar neutrino experiments, the shape of Super-Kamiokande recoil electron energy spectrum and the results of reactor experiments at Bugey (MUNU) and Rovno [88]. As statistical procedure a Bayesian method is used with flat priors and minimization of χ^2 with respect to θ_{12}, $\Delta m_\odot^2$, $|\Lambda_1|$, $|\Lambda_2|$, δ', in order to extract a probability distribution for $|\mathbf{\Lambda}|$. The details are given in [91], where the 90% CL bounds

$$|\mathbf{\Lambda}| < \begin{cases} 6.3 \times 10^{-10} \mu_B \text{ (solar data)}, \\ 2.0 \times 10^{-10} \mu_B \text{ (solar + reactor data)} \end{cases} \tag{99}$$

are presented.

Some comments are at order [91]:

▷ We want to stress once more that the bounds in (99) apply to $|\mathbf{\Lambda}|^2 = |\lambda_{e\mu}|^2 + |\lambda_{\mu\tau}|^2 + |\lambda_{\tau e}|^2 = \sum_{j<k} |\lambda_{jk}|^2$. Therefore, all transition moments are bounded in a basis-independent way.
▷ In (96), in the limit $P_{e1}^S \to 0$, the quantity $|\Lambda_2|$ drops out of the effective MM. Therefore, for small P_{e1}^S, solar neutrino data become less stringent for $|\Lambda_2|$ and thus for $|\mathbf{\Lambda}|$ as well. This is the case for Super-Kamiokande data (see the first bound in (99)) where P_{e1}^S is small due to the relatively high neutrino energy (see (53)).
▷ Reactor data give a good bound on $|\Lambda_2|$ and are, therefore, complementary to present solar neutrino data.
▷ The BOREXINO experiment [92] could improve the second bound in (99) by nearly one order of magnitude because it is sensitive to relatively low neutrino energies where $P_{e1}^S \sim 0.5$.

4   Models for Neutrino Masses and Mixing

4.1   Introduction and Scope

For a start, we present the problems of model building for neutrino masses and mixing by asking appropriate questions and formulating answers, if available.

1. Can neutrino masses and mixing be *accommodated* in a model?
   This is no problem; the simplest possibility is given by the SM + 3 right-handed neutrino singlets $\nu_R + L$ conservation which allows to accommodate arbitrary masses and mixing of Dirac neutrinos in complete analogy to the quark sector.
2. Why are neutrino masses much smaller than the charged lepton masses?
   There are two popular proposals for a solution to this problem:
   • Radiative neutrino masses;
   • Seesaw mechanism [93,94].

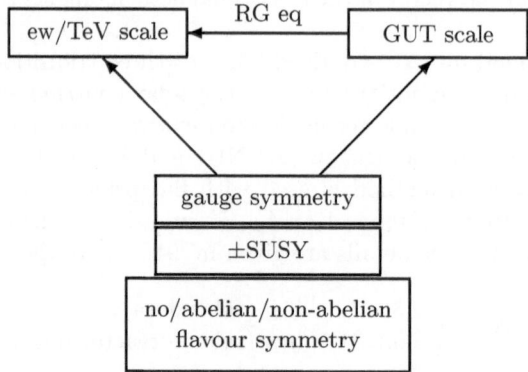

Fig. 8. Symbolical presentation of the variety of possibilities for model building. The upper two blocks show the two energy scales at which the symmetries, indicated in the three blocks below, could apply. RG eq stands for renormalization group equation, ± for with/without.

3. Can one reproduce the special features of ν masses and mixing?
   Let us list the features one would like to explain:

   F1 $\theta_\odot \sim 34°$ (large but non-maximal);
   F2 $\theta_{\mathrm{atm}} \simeq 45°$;
   F3 $|U_{e3}|^2 \equiv s_{13}^2 \lesssim 0.05$;
   F4 $\Delta m_\odot^2 / \Delta m_{\mathrm{atm}}^2 \sim 0.03$.

   We have used the obvious notation where the subscripts $\odot$ and atm refer to solar and atmospheric neutrinos, respectively.

   To explain the listed features is the most difficult and largely unsolved task. There are myriads of textures or models – for reviews see [95]. One of the problems is that there are still not enough clues where to start model building. Symbolically, the difficulties of model building are presented in Fig. 8. If the symmetries apply at the GUT scale, the renormalization group equation transports relations from the GUT scale to the low scale; this transport could contribute to generate some of the desired features – for a review see [96]. It is completely unknown if the explanation of the features F1-4 is independent of general fermion mass problem or not. In the following we will assume independence.
   The scope of the following sections is to discuss some simple extensions of the SM by addition of scalar multiplets and of right-handed neutrino singlets ν_R. All these extensions will yield Majorana neutrino masses. We will only discuss the lepton sector.

4.2   The Standard Model with Additional Scalar Multiplets

In the SM there are only two types of lepton multiplets:

$$SU(2)\times\ U(1)$$
$$D_L \quad \tfrac{1}{2} \quad Y = -1 \text{ left-handed doublets}$$
$$\ell_R \quad \underline{0} \quad Y = -2 \text{ right-handed singlets}$$

Here Y is the hypercharge and the underlined numbers indicate the weak isospin of the multiplet.

By forming all possible leptonic bilinears one obtains the possible scalar gauge multiplets which can couple to fermions [97]:

$$\bar{D}_L \otimes \ell_R \;\tfrac{1}{2} \otimes \underline{0} = \tfrac{1}{2} \quad Y = -1 \quad \phi \quad \text{doublet}$$
$$D_L \otimes D_L \;\tfrac{1}{2} \otimes \tfrac{1}{2} = \underline{0} \oplus \underline{1}\; Y = -2 \begin{cases} \eta^+ & \text{singlet} \\ \Delta & \text{triplet} \end{cases}$$
$$\ell_R \otimes \ell_R \quad \underline{0} \otimes \underline{0} = \underline{0} \quad Y = -4 \quad k^{++} \quad \text{singlet}$$

The hypercharge in this table refers to the leptonic bilinear. In the following we will discuss extensions of the SM with these scalar multiplets (except the trivial extension where only Higgs doublets are added).

The Zee model: This model [98,99] is defined as SM with $2\phi + \eta^+$ and, therefore, has the Lagrangian

$$\mathcal{L} = \cdots + \left[f_{\alpha\beta} D_{\alpha L}^T C^{-1} i\tau_2 D_{\beta L}\, \eta^+ - \mu\, \phi_1^\dagger \tilde{\phi}_2 \eta^+ + \text{H.c.} \right], \qquad (100)$$

where the dots indicate the SM part and terms of the Higgs potential which are not interesting for the following discussion. Note the antisymmetry of the coupling matrix: $f_{\alpha\beta} = -f_{\beta\alpha}$.

Since no fermionic multiplets have been added to the SM, the ensuing neutrino masses can only be of Majorana type. Thus the total lepton number L must be broken, otherwise neutrino masses will be strictly forbidden. Let us assign a lepton number to η^+ from its Yukawa couplings in (100). Then we have the following list of lepton numbers of the multiplets of the Zee model:

$$\begin{array}{c|cccc} & D_L & \ell_R & \phi_{1,2} & \eta^+ \\ \hline L & 1 & 1 & 0 & -2 \end{array} \qquad (101)$$

Thus, L is indeed explicitly violated by the term $\phi_1^\dagger \tilde{\phi}_2 \eta^+$ in the Lagrangian. Such a term can only be formed if *two* Higgs doublets are present (if $\phi_1 = \phi_2 = \phi$, then $\phi^\dagger \tilde{\phi} \equiv 0$). In the Zee model a neutrino mass matrix $\mathcal{M}_\nu$ appears at the 1-loop level.

A lot of work has been done on the restricted Zee model [100] where only one Higgs doublet, say ϕ_1, couples to the leptons. In this particularly simple version one has $\mathcal{M}_\nu \propto \left((m_\alpha^2 - m_\beta^2) f_{\alpha\beta} \right)$, where $m_{\alpha,\beta}$ denote the charged lepton masses. Note that without loss of generality we have chosen a basis where the charged lepton mass matrix is diagonal. The restricted Zee model is practically ruled out now because it allows only maximal solar mixing [101]; furthermore, it requires serious fine-tuning, namely $m_\tau^2 |f_{e\tau}| \simeq m_\mu^2 |f_{e\mu}|$; finally, the smallness of the neutrino masses has to be achieved by $|f_{\alpha\beta}| \lesssim 10^{-4}$. If both $\phi_{1,2}$ couple to the leptons, then non-maximal solar mixing is admitted [102], the fine-tuning problem is somewhat alleviated, but the third point remains. For further recent literature on the Zee model see, e.g. [103]

The Zee-Babu model: This model is defined as [99,104] SM $+ \eta^+ + k^{++}$ and has the Lagrangian

$$
\begin{aligned}
\mathcal{L} = \cdots + \big[& f_{\alpha\beta} D_{\alpha L}^T C^{-1} i\tau_2 D_{\beta L}\, \eta^+ \\
& + h_{\alpha\beta} \ell_{\alpha R}^T C^{-1} \ell_{\beta R}\, k^{++} - \mu \eta^- \eta^- k^{++} + \text{H.c.} \big]
\end{aligned}
\tag{102}
$$

with a symmetric coupling matrix $h_{\alpha\beta} = h_{\beta\alpha}$. In addition to the assignments (101), we have $L(k^{++}) = -2$. Again, L must be explicitly broken, which is now achieved by the μ-term in (102).

Since in this model there is only one Higgs doublet, L is still conserved at the 1-loop level (see previous section on the Zee model) and the neutrino mass matrix appears at the 2-loop level: $\mathcal{M}_\nu \propto \tilde{f} \hat{m}_\ell \tilde{h}^* \hat{m}_\ell \tilde{f}$ with $\tilde{f} = (f_{\alpha\beta})$, $\tilde{h} = (h_{\alpha\beta})$ and $\hat{m}_\ell = \text{diag}\,(m_e, m_\mu, m_\tau)$. This model has the following properties [105]. Since $\tilde{f}$ is antisymmetric, the lightest neutrino mass is zero; the solar LMA MSW solution and, e.g., a hierarchical neutrino mass spectrum require the fine-tuning $|h_{\mu\mu}| : |h_{\mu\tau}| : |h_{\tau\tau}| \simeq 1 : (m_\mu/m_\tau) : (m_\mu/m_\tau)^2$; all scalar masses are in the TeV range; in order to reproduce the neutrino masses inferred from atmospheric and solar data one needs $|f_{\alpha\beta}|, |h_{\alpha\beta}| \lesssim 0.1$, thus neutrino masses are naturally small as a consequence of the 2-loop mechanism; finally, rare decays like $\tau \to 3\mu$, $\mu \to e\gamma$ should be within reach of forthcoming experiments.

We note that from the Zee model and Zee-Babu model we have seen that models with radiative neutrino mass generation are prone to excessive fine-tuning because the hierarchy of the charged lepton masses works against the features needed in the neutrino sector.

The triplet model: This model is defined by adding a scalar triplet Δ to the SM and has the Lagrangian

$$
\begin{aligned}
\mathcal{L} = \cdots + \big[& \tfrac{1}{2} g_{\alpha\beta} D_{\alpha L}^T C^{-1} i\tau_2 \Delta D_{\beta L} + \text{H.c.} \big] \\
& - \tfrac{1}{2} M^2 \text{Tr}\, \Delta^\dagger \Delta - \big(\mu \, \phi^\dagger \Delta \tilde{\phi} + \text{H.c.} \big)
\end{aligned}
\tag{103}
$$

The electric charge eigenfields and the VEV of the triplet are given as

$$
\Delta = \begin{pmatrix} H^+ & \sqrt{2} H^{++} \\ \sqrt{2} H^0 & -H^+ \end{pmatrix}, \quad \langle H^0 \rangle_0 = \frac{1}{\sqrt{2}} v_T,
\tag{104}
$$

respectively. Note that $g_{\alpha\beta} = g_{\beta\alpha}$. As in the previous two models, we can make the assignment $L(\Delta) = -2$, then L is explicitly broken by the μ-term in (103). The original model without the μ-term and spontaneous L breaking [106] is ruled out by the non-discovery of the Goldstone boson and a light scalar at LEP. This model leads to the tree-level neutrino mass matrix $\mathcal{M}_\nu = v_T(g_{\alpha\beta})$. The LEP data require $|v_T/v| \lesssim 0.03$ [107], where v is the VEV of the SM Higgs doublet. However, if the coupling constants $g_{\alpha\beta}$ are about $0.01 \div 0.1$, then in order to obtain small neutrino masses the triplet VEV must be much smaller, namely $v_T \sim 0.1 \div 1$ eV. There are two ways to get a small v_T: Firstly, one can assume that $M, |\mu| \gg v$, then $|v_T| \simeq |\mu| v^2/M^2$ (scalar or type II seesaw mechanism) [94,109]; secondly, with $M \sim v, |\mu| \ll v$ one has $|v_T| \sim |\mu|$ [108]. In some sense the mechanism for obtaining small neutrino masses in the triplet model is analogous to the seesaw mechanism [93,94] (see subsequent subsection) since in both cases one has to introduce a second scale much larger (smaller) than the electroweak scale in order to generate small neutrino masses.

4.3   The Seesaw Mechanism

The seesaw mechanism [93,94] is implemented in the simplest way in the SM $+ 3\nu_R + L$ violation. This is primarily an extension in the fermion sector of the SM. Note that one could also choose two or more than three right-handed singlets ν_R. The starting point is the Lagrangian

$$\mathcal{L} = \cdots - \left[\sum_j \left(\bar{\ell}_R \phi_j^\dagger \Gamma_j + \bar{\nu}_R \tilde{\phi}_j^\dagger \Delta_j\right) D_L + \text{H.c.}\right] \\ + \left(\tfrac{1}{2} \nu_R^T C^{-1} M_R^* \nu_R + \text{H.c.}\right), \tag{105}$$

where the mass matrix M_R of the right-handed singlets must be symmetric due to their fermionic anticommutation property. The number n_H of Higgs doublets is irrelevant for the seesaw mechanism. Defining the mass matrices

$$M_\ell = \frac{1}{\sqrt{2}} \sum_j v_j^* \Gamma_j, \quad M_D = \frac{1}{\sqrt{2}} \sum_j v_j \Delta_j, \tag{106}$$

where M_ℓ is the mass of the charged leptons and M_D is the so-called "Dirac mass matrix" for the neutrinos, the total *Majorana* mass matrix for all six left-handed fields ω_L is given by [110]

$$\mathcal{M}_{D+M} = \begin{pmatrix} 0 & M_D^T \\ M_D & M_R \end{pmatrix} \quad \text{with} \quad \omega_L = \begin{pmatrix} \nu_L \\ C(\bar{\nu}_R)^T \end{pmatrix}. \tag{107}$$

The VEVs v_j fulfill $v = \sqrt{|v_1|^2 + \ldots |v_{n_H}|^2} = (\sqrt{2}\, G_F)^{-1/2} \simeq 246$ GeV.

The basic assumption for implementing the seesaw mechanism is $m_D \ll m_R$, where $m_{D,R}$ are the scales of $M_{D,R}$, respectively. With this assumption, one obtains the mass matrix of the light neutrinos

$$\mathcal{M}_\nu = -M_D^T M_R^{-1} M_D, \tag{108}$$

which is valid up to corrections of order $(m_D/m_R)^2$. The mass matrix of heavy neutrinos is given by $\mathcal{M}_\nu^{\text{heavy}} = M_R$. Diagonalizing the mass matrices by

$$(U_R^\ell)^\dagger M_\ell U_L^\ell = \hat{m}_\ell, \quad V^T \mathcal{M}_\nu V = \hat{m}, \tag{109}$$

the neutrino mixing matrix for the light neutrinos is given by

$$U_M = (U_L^\ell)^\dagger V. \tag{110}$$

As discussed earlier, phases multiplying U_M from the left are unphysical because they can be absorbed into the charged lepton fields.

The seesaw mechanism contains three sources for neutrino mixing: M_ℓ, M_D and M_R. Therefore, it is a rich playground for model building. It can also be combined with radiative neutrino mass generation – for an example see next subsection. Let us discuss the order of magnitude of the scale m_R. If we choose as a typical neutrino mass $m_\nu \sim \sqrt{\Delta m_{\text{atm}}^2} \sim 0.05$ eV and assume $m_D \sim m_{\mu,\tau}$, we obtain $m_R \sim 10^8 \div 10^{11}$ GeV. On the other hand, if m_D is of the order of the electroweak scale, then $m_R \sim 10^{15}$ GeV and there could be a possibility to identify it with the GUT scale.

4.4 Combining the Seesaw Mechanism with Radiative Mass Generation

Let us now consider the SM with *two* Higgs doublets $\phi_{1,2}$, add a *single* field ν_R and allow for L violation [111,112]. We do not employ any flavour symmetry.

In this case the Yukawa coupling matrices $\Delta_{1,2}$ and M_D are 1×3 matrices (vectors!). Here we give a precise meaning to the scale m_D and identify it with the length of the vector M_D. L violation is induced by the Majorana mass term with mass $M_R \gg m_D$ of the neutrino singlet ν_R. Then at the tree level the seesaw mechanism is operative:

$$\text{Tree level:} \quad m_1 = m_2 = 0, \quad m_3 \simeq m_D^2/M_R. \tag{111}$$

That two masses are zero at the tree level is a consequence of M_D being a 1×3 matrix, which, therefore, maps two vectors onto 0; this feature is then operative in $\mathcal{M}_{D+M}$ (107) and $\mathcal{M}_\nu$ (108).

At 1-loop level, m_2 becomes non-zero by neutral-scalar exchange, bit m_1 remains zero:

$$\text{1-loop:} \quad m_1 = 0, \quad m_2 \sim \frac{1}{16\pi^2} m_3 \frac{M_0^2}{v^2} \ln(M_R/M_0)^2. \tag{112}$$

In this order-of-magnitude relation for m_2 [112], the mass $M_0 \sim v$ is a typical scalar mass (there are three physical neutral scalars in this model). A general discussion for arbitrary numbers of left-handed lepton doublets, right-handed neutrinos singlets and Higgs doublets is found in [111]. The full calculation of the dominant 1-loop corrections to the seesaw mechanism is presented in [113].

The model has the following properties. It predicts a hierarchical spectrum, therefore $\sqrt{\Delta m_\odot^2/\Delta m_{\text{atm}}^2} \simeq m_2/m_3 \overset{\text{exp.}}{\sim} 0.17$. It gives the correct order of magnitude of m_2/m_3 with $1/16\pi^2 \simeq 0.0063$, $M_0^2/v^2 \sim 1$, $\ln(M_R/M_0) \sim 10$. The most interesting property of these 1-loop corrections to the seesaw mechanism is the fact that the suppression relative to the tree level terms is given solely by the loop integral factor $1/16\pi^2$ [111,113]. The mixing angles are undetermined, but without fine-tuning they will be large in general. Consequently, the model has no argument for $\theta_{\text{atm}} \simeq 45°$ and small angle θ_{13}; these must be reproduced by tuning the parameters of the model. This is easily achieved because in good approximation $|U_{\alpha 3}| = |M_{D\alpha}|/m_D$ [112].

It is interesting to note that R-parity-violating supersymmetric models have a built-in seesaw mechanism where the heavy Majorana neutrinos are replaced by the neutralinos which are also Majorana particles. See, e.g., [114] and citations therein.

4.5   A Model for Maximal Atmospheric Neutrino Mixing

Here we want to discuss a model based on tree-level seesaw masses. While in the previous model the emphasis was on explaining the ratio $\Delta m_\odot^2/\Delta m_{\text{atm}}^2$ and we assumed to obtain the mixing angles by tuning of model parameters, here we take the opposite attitude. One of the problems of obtaining maximal atmospheric neutrino mixing and large but non-maximal solar neutrino mixing by a symmetry is that the mixing matrix (110) has a contribution also from the charged lepton sector. Now we introduce a framework where this problem is avoided by having M_R as the only source of neutrino mixing.

The framework: We start with the Lagrangian (105) and allow for an arbitrary number n_H of Higgs doublets. Then how can one avoid flavour-changing interactions via tree-level scalar interactions? We assume that the family lepton numbers $L_{e,\mu,\tau}$ are conserved in all terms of dimension 4 in the Lagrangian but L_α *is softly broken by the ν_R mass term.* This allows one to kill several birds with one stone: Flavour-changing neutral interactions are forbidden at the tree level, M_ℓ and M_D are diagonal and neutrino mixing stems exclusively from M_R [115], as announced above. However, soft breaking of lepton numbers occurs at the *high* scale m_R. It has been shown in [116] that this assumption yields a perfectly viable theory with interesting properties: All 1-loop flavour-changing vertices are finite because of soft L_α breaking; for $n_H > 1$, such vertices where the boson leg is a neutral scalar and the exchanged boson is a charged scalar do not decouple in the limit $m_R \to \infty$ (for $n_H = 1$ there is decoupling); diagrams with a γ or a Z and box diagrams always decouple. As a consequence, the amplitudes of $\mu \to e\gamma$, $Z \to e^-\mu^+, \ldots$ are suppressed by $1/m_R^2$, whereas, e.g., the amplitude of $\mu \to 3e$ tends to a constant for $m_R \to \infty$ and is suppressed – though much less than the previous amplitudes – because it contains a product of four Yukawa couplings. The decay rate of the latter process in this framework might be within reach of forthcoming experiments. For details see [116].

Maximal atmospheric neutrino mixing: Within the framework of soft L_α-breaking we introduce now a Z_2 symmetry:

$$Z_2 : \quad D_{\mu L} \leftrightarrow D_{\tau L}, \, \nu_{\mu R} \leftrightarrow \nu_{\tau R}, \, \mu_R \leftrightarrow \tau_R. \tag{113}$$

This symmetry makes M_D and M_R Z_2-invariant and, therefore, transfers to the neutrino mass matrix $\mathcal{M}_\nu$. Thus we obtain the result

$$\mathcal{M}_\nu = \begin{pmatrix} x & y & y \\ y & z & w \\ y & w & z \end{pmatrix} \Rightarrow U = \begin{pmatrix} \cos\theta & \sin\theta & 0 \\ -\frac{\sin\theta}{\sqrt{2}} & \frac{\cos\theta}{\sqrt{2}} & \frac{1}{\sqrt{2}} \\ \frac{\sin\theta}{\sqrt{2}} & -\frac{\cos\theta}{\sqrt{2}} & \frac{1}{\sqrt{2}} \end{pmatrix}, \tag{114}$$

where $\theta \equiv \theta_{12}$ is the solar mixing angle. Summarizing, we have constructed a model where the solar mixing angle is free and without fine-tuning it will be large but non-maximal; atmospheric mixing is maximal; $|U_{e3}| = s_{13} = 0$.

The above results are stable under radiative corrections because the mass matrix (114) was realized by the symmetries $U(1)_{L_e} \times U(1)_{L_\mu} \times U(1)_{L_\tau}$, which are softly broken, and Z_2, which has to be broken spontaneously in order to achieve $m_\mu \neq m_\tau$. This can be done with a minimum of three Higgs doublets and an auxiliary Z_2', without destroying the form of the mass matrix (114) [115]. Since the Z_2 of (113) does not commute with $L_{\mu,\tau}$, the full group generated by our symmetries is non-abelian [117]. The essential features of this model can be embedded in an $SU(5)$ GUT [117].

5   Conclusions

In recent years we have witnessed great progress in neutrino physics. Eventually, it has been confirmed that the solar neutrino puzzle is solved by neutrino oscillations, first conceived by Bruno Pontecorvo in 1957. At the same time, matter effects in neutrino oscillations – as occurring in the LMA MSW solution – have turned out to play a decisive role. It is firmly believed that also the atmospheric neutrino deficit problem is solved by neutrino oscillations though for the time being the final proof is still missing but is expected to be provided soon by LBL experiments.

Despite the great progress, which provides us with a first glimpse beyond the SM, there are still many things we would like to know. For instance, Are neutrinos Dirac or Majorana particles? What are the absolute neutrino masses? Of what type is the neutrino mass spectrum? Do neutrinos have sizeable magnetic moments?

As for field-theoretical models of neutrino masses and mixing, theorists are groping in the dark. As it has turned out, neutrino mass spectra and the neutrino mixing matrix are very different from the quark sector. Though there are many ideas, very few of them account naturally for some of the neutrino properties. Despite the big increase in our knowledge about neutrinos even basic questions for model building have no answer at the moment. Some of the basic questions are the following: Is the neutrino mass and mixing problem independent of the general fermion mass problem? Are neutrino masses small by radiative, seesaw or other mechanisms? Is the solution for the neutrino mass and mixing problem situated at the TeV scale or the GUT scale? Do we need a flavour symmetry for its solution and, if yes, of what type is this symmetry?

We want to stress that it is no problem to *accommodate* neutrino masses and mixing in theories, but the problem is to *explain* the specific features for neutrinos. As for the neutrino nature, there is a theoretical bias toward Majorana nature, e.g. from the seesaw mechanism and GUTs, in particular, from GUTs based on $SO(10)$. For the time being, there are simply not enough clues for a definite mechanism for neutrino masses and mixing. Among others, possible future clues provided by experiment would be an atmospheric mixing angle very close to $45°$, which would point toward a non-abelian flavour symmetry; knowledge of the value of $|U_{e3}| = s_{13}$ and the neutrino mass spectrum, which would teach us more about the mass matrix; discovery of scalars with masses $\lesssim 1$ TeV, which would show that fermion masses are most probably generated by the Higgs mechanism; discovery of SUSY partners of ordinary particles, which would assure us that we have to take SUSY into account in model building. Evidence for flavour-changing decays like $\mu \to e\gamma$, $\mu \to 3e$, etc. would also give a valuable input.

At any rate, ongoing and future experiments will continue to provide us with exciting results, which will enhance the prospects of constructing viable mechanisms for explaining the specific neutrino features.

Acknowledgements

I wish to thank the Organizing Committee for the invitation to Schladming and A. Bartl, G. Ecker, M. Hirsch and H. Neufeld for discussions.

References

1. A.Yu. Smirnov, Talk given at the *International Workshop NOON2003*, February 10–14, 2003, Kanazawa, Japan, hep-ph/0306075; Talk given at the *11th Workshop on Neutrino Telescopes*, Venice, March 11–14, 2003, hep-ph/0305106.
2. S. Goswami, Pramana **60**, 261 (2003) [hep-ph/0305111].
3. C. Giunti, hep-ph/0305139.

4. G. Drexlin, Contribution to these proceedings.
5. J.N. Bahcall, S. Basu and M.P. Pinsonneault, Phys. Lett. B **433**, 1 (1998); Astrophys. J. **555**, 990 (2001).
6. SNO Collaboration, Q.R. Ahmad et al., Phys. Rev. Lett. **89**, 011301 (2002) [nucl-ex/0204008].
7. KamLAND Collaboration, K. Eguchi et al., Phys. Rev. Lett. **90**, 021802 (2003) [hep-ex/0212021].
8. S.M. Bilenky and B. Pontecorvo, Phys. Rep. **41**, 225 (1978).
9. S.M. Bilenky, hep-ph/9908335.
10. L.B. Okun, M.G. Schepkin and I.S. Tsukerman, Nucl. Phys. B **650**, 443 (2003); Err. *ibid.* **656**, 255 (2003) [hep-ph/0211241].
11. S.M. Bilenky and S.T. Petcov, Rev. Mod. Phys. **59**, 671 (1987).
12. S.M. Bilenky, C. Giunti and W. Grimus, Prog. Part. Nucl. Phys. **43**, 1 (1999) [hep-ph/9812360].
13. P.B. Pal, Int. J. Mod. Phys. A **7**, 5387 (1992);
    K. Zuber, Phys. Rept. **305**, 295 (1998);
    P. Fisher, B. Kayser, and K. S. McFarland, Ann. Rev. Nucl. Part. Sci. **49**, edited by C. Quigg, V. Luth, and P. Paul (Annual Reviews, Palo Alto, California, 1999), p. 481;
    M. Gonzalez-Garcia and Y. Nir, Rev. Mod. Phys. **75**, 345 (2003) [hep-ph/0202058].
14. R.N. Mohapatra and P.B. Pal, *Massive Neutrinos in Physics and Astrophysics*, World Scientific Lecture Notes in Physics, Vol. 41 (World Scientific, 1991);
    C.W. Kim and A. Pevsner, *Neutrinos in Physics and Astrophysics*, Contemporary Concepts in in Physics, Vol. 8 (Academic Press, 1993).
15. L. Wolfenstein, Phys. Rev. D **17**, 2369 (1978); *ibid.* **20**, 2634 (1979).
16. S.P. Mikheyev and A.Yu. Smirnov, Yad. Fiz. **42**, 1441 (1985) [Sov. J. Nucl. Phys. **42**, 913 (1985)]; Il Nuovo Cim. C **9**, 17 (1986).
17. LSND Collaboration, A. Aguilar et al., Phys. Rev. D **64**, 112007 (2001) [hep-ex/0104049].
18. M. Maltoni, T. Schwetz, M.A. Tórtola and J.W.F. Valle, Nucl. Phys. B **643**, 321 (2002) [hep-ph/0207157]; in Proc. of *30th International Meeting on Fundamental Physics (IMFP2002)*, Jaca, Spain, January 28–February 1, 2002, Nucl. Phys. B (Proc. Suppl.) **114**, 203 (2003) [hep-ph/0209368].
19. M. Sorel, J. Conrad and M. Shaevitz, hep-ph/0305255.
20. CHOOZ Experiment, M. Apollonio et al., Eur. Phys. J. C **27**, 331 (2003) [hep-ex/0301017].
21. Palo Verde Experiment, F. Boehm et al., Phys. Rev. D **64**, 112001 (2001) [hep-ex/0107009].
22. M. Zrałek, Acta Phys. Pol. B **29**, 3925 (1998) [hep-ph/9810543].
23. M. Beuthe, Phys. Rept. **375**, 105 (2003) [hep-ph/0109119].
24. J. Rich, Phys. Rev. D **48**, 4318 (1993).
25. M. Beuthe, Phys. Rev. D **66**, 013003 (2002) [hep-ph/0202068].
26. C. Giunti, C.W. Kim and J.A. Lee, Phys. Rev. D **48**, 4310 (1993) [hep-ph/9305276].
27. W. Grimus and P. Stockinger, Phys. Rev. D **54**, 3414 (1996) [hep-ph/9603430].
28. H. Lipkin, Phys. Lett. B **348**, 304 (1995) [hep-ph/9501269].
29. W. Grimus, P. Stockinger and S. Mohanty, Phys. Rev. D **59**, 013011 (1999) [hep-ph/9807442].

30. H. Lipkin, hep-ph/0304187.
31. L. Stodolsky, Phys. Rev. D **58**, 036006 (1998) [hep-ph/9802387].
32. B. Kayser, Phys. Rev. D **24**, 110 (1981).
33. W. Grimus, S. Mohanty P. and Stockinger, Phys. Rev. D **61**, 033001 (2000) [hep-ph/9904285].
34. K. Kiers, S. Nussinov and N. Weiss, Phys. Rev. **53**, 537 (1996) [hep-ph/9506271].
35. C. Giunti, hep-ph/0302026.
36. S.P. Mikheyev and A.Yu. Smirnov, Prog. Part. Nucl. Phys. **23**, 41 (1989).
37. H.A. Bethe, Phys. Rev. Lett. **56**, 1305 (1986).
38. J.J. Sakurai, *Advanced Quantum Mechanics* (Addison-Wesley, Reading, Massachusetts, 1967), p. 62.
39. J. Liu, Phys. Rev. D **45**, 1428 (1992).
40. W. Grimus and T. Scharnagl, Mod. Phys. Lett. A **8**, 1943 (1993); B. Bekman, J. Gluza, J. Holeczek, J. Syska and M. Zrałek, Phys. Rev. D **66**, 093004 (2002) [hep-ph/0207015].
41. M.C. Gonzalez-Garcia and C. Peña-Garay, in Proc. of *XIXth International Conference on Neutrino Physics and Astrophysics*, Sudbury, Canada, 16-21 June 2000, Nucl. Phys. B (Proc. Suppl.) **91**, 80 (2001) [hep-ph/0009041].
42. S. Choubey, A. Bandyopadhyay, S. Goswami and D.P. Roy, Talk given at Conference on *Physics beyond the Standard Model: Beyond the Desert 02*, Oulu, Finland, 2–7 June 2002, hep-ph/0209222.
43. L.I. Schiff, *Quantum Mechanics* (McGraw-Hill, 1987).
44. W.C. Haxton, Phys. Rev. Lett. **57**, 1271 (1986).
45. S.J. Parke, Phys. Rev. Lett. **57**, 1275 (1986).
46. T.K. Kuo and J. Pantaleone, Rev. Mod. Phys. **61**, 937 (1989).
47. S.P. Mikheyev and A.Yu. Smirnov, Sov. Phys. JETP **65**, 230 (1987).
48. A. Friedland, Phys. Rev. D **64**, 013008 (2001) [hep-ph/0010231].
49. J.N. Bahcall, M.C. Gonzalez-Garcia and C. Peña-Garay, JHEP **0207**, 054 (2002) [hep-ph/0204314].
50. Cheng-Wei Chiang and L. Wolfenstein, Phys. Rev. D **63** 057303 (2001) [hep-ph/0010213].
51. S.M. Bilenky, C. Giunti, J.A. Grifols and E. Masso, Phys. Rept. **379**, 69 (2003) [hep-ph/0211462].
52. Particle Data Group, K. Hagiwara et al., Phys. Rev. D **66**, 010001 (2002).
53. G.L. Fogli et al., Phys. Rev. D **66**, 093008 (2002) [hep-ph/0208026]; M.C. Gonzalez-Garcia and C. Peña-Garay, hep-ph/0306001.
54. M. Shiozawa, Talk given at the *XXth International Neutrino Conference on Neutrino Physics and Astrophysics*, Munich, May 25–30, 2002, Transparencies at http://neutrino2002.ph.tum.de.
55. T. Nakaya, in Proc. of *XXth International Neutrino Conference on Neutrino Physics and Astrophysics*, Munich, May 25–30, 2002, Nucl. Phys. B (Proc. Suppl.) **118**, 210 (2003).
56. S. Geer, Phys. Rev. D **57**, 6989 (1998); Err. *ibid.* **59**, 039903 (1999) [hep-ph/9712290].
57. V. Barger, S. Geer, R. Raja and K. Whisnant, Phys. Rev. D **62**, 013004 (2000) [hep-ph/9911524].
58. M. Freund, M. Lindner, S.T. Petcov and A. Romanino, Nucl. Phys. B **578**, 27 (2000) [hep-ph/9912457].

59. P. Huber, M. Lindner and W. Winter, Nucl. Phys. B **645**, 3 (2002) [hep-ph/0204352].
60. M. Lindner, in Proc. of *XXth International Neutrino Conference on Neutrino Physics and Astrophysics*, Munich, May 25–30, 2002, Nucl. Phys. B (Proc. Suppl.) **118**, 199 (2003) [hep-ph/0210377].
61. I. Schur, Am. J. Math. **67**, 472 (1945);
    see also B. Zumino, J. Math. Phys. **3**, 1055 (1962);
    W. Grimus and G. Ecker, J. Phys. A: Math. Gen. **21**, 2825 (1988).
62. L. Wolfenstein, Phys. Lett. **107B**, 77 (1981);
    S.M. Bilenky, N.P. Nedelcheva and S.T. Petcov, Nucl. Phys. B **247**, 61 (1984);
    B. Kayser, Phys. Rev. D **30**, 1023 (1984).
63. M. Fukugita and T. Yanagida, Phys. Lett. B **174**, 45 (1986).
64. A. Pilaftsis, Int. J. Mod. Phys. A **14**, 1811 (1999) [hep-ph/9812256];
    W. Buchmüller and M. Plümacher, Int. J. Mod. Phys. A **15**, 5047 (2000) [hep-ph/0007176];
    W. Buchmüller, Lectures given at the *2001 European School of High-Energy Physics*, Beatenberg, Switzerland, hep-ph/0204288.
65. J.A. Casas and A. Ibarra, Nucl. Phys. B **618**, 171 (2002) [hep-ph/0103065].
66. G.C. Branco, T. Morozumi, B.M. Nobre and M.N. Rebelo, Nucl. Phys. B **617**, 475 (2001) [hep-ph/0107164];
    M.N. Rebelo, Phys. Rev. D **67**, 013008 (2003) [hep-ph/0207236]
67. S. Pascoli, S.T. Petcov and W. Rodejohann, hep-ph/0302054.
68. V.I. Tretyak and Y.G. Zdesenko, Atomic Data and Nuclear Data Tables **80**, 83 (2002).
69. Heidelberg-Moscow Collaboration, H.V. Klapdor-Kleingrothaus et al., Eur. Phys. J. A **12**, 147 (2001).
70. S.R. Elliot and P. Vogel, Ann. Rev. Nucl. Sci. **52**, 115 (2002) [hep-ph/0202264].
71. R.N. Mohapatra, in Proc. of *XVIIIth International Neutrino Conference on Neutrino Physics and Astrophysics*, Takayama, Japan, June 4–9, 1998, Nucl. Phys. B (Proc. Suppl.) **77**, 376 (1999) [hep-ph/9808284].
72. IGEX Collaboration, C.E. Aalseth et al., Phys. Rev. D **65**, 092007 (2002) [hep-ex/0202026].
73. R.N. Mohapatra, Phys. Rev. D **34**, 3457 (1986);
    M. Hirsch, H.V. Klapdor-Kleingrothaus and S.G. Kovalenko, Phys. Rev. Lett. **75**, 17 (1995).
74. M.B. Smy (Super-Kamiokande Collaboration), in Proc. of *XXth International Neutrino Conference on Neutrino Physics and Astrophysics*, Munich, May 25–30, 2002, Nucl. Phys. B (Proc. Suppl.) **118**, 199 (2003) [hep-ex/0208004].
75. J. Schechter and J.W.M. Valle, Phys. Rev. D **25**, 2951 (1982).
76. E. Takasugi, Phys. Lett. **149B**, 372 (1984).
77. H. Päs and T.J. Weiler, Phys. Rev. D **63**, 113015 (2001) [hep-ph/0101091];
    Talk presented at *SUSY02, 10th International Conference on Supersymmetry and Unification of Fundamental Interactions*, DESY, Hamburg, June 17–23, 2002, hep-ph/0212194.
78. Y. Farzan and A.Yu. Smirnov, Phys. Lett. B **557**, 224 (2003) [hep-ph/0211341].
79. S. Pascoli and S.T. Petcov, Phys. Lett. B **544**, 239 (2002) [hep-ph/0205022].
80. S.M. Bilenky, C. Giunti, W. Grimus, B. Kayser and S.T.Petcov, Phys. Lett. B **465**, 193 (1999) [hep-ph/9907234].

81. M. Czakon, J. Gluza and M. Zrałek, Acta Phys. Pol. B **30**, 3121 (1999) [hep-ph/9910357]; hep-ph/0003161.
82. M. Colless et al., MNRAS **328**, 1039 (2001) [astro-ph/0106498].
83. D.N. Spergel et al., astro-ph/0302209.
84. G.G. Raffelt and W. Rodejohann, *Massive neutrinos in Astrophysics*, Lectures given at *4th National Summer School "Grundlagen und neue Methoden der theoretischen Physik"*, August 31–September 11, 1998, Saalburg, Germany, hep-ph/9912397.
85. S. Sarkar, Invited Contribution to a special issue of the Proc. of the Indian National Academy of Sciences, hep-ph/0302175;
    S. Pastor, Talk given at the *Xth International Workshop on Neutrino Telescopes*, Venice, March 11–24, 2003, hep-ph/0306233.
86. W. Grimus and T. Schwetz, Nucl. Phys. B **587**, 45 (2000) [hep-ph/0006028].
87. D.Y. Bardin, S.M. Bilenky and B. Pontecorvo, Phys. Lett. **32B**, 68 (1970);
    A.V. Kyuldjiev, Nucl. Phys. B **243**, 387 (1984).
88. Rovno reactor, A.I. Derbin et al., JETP Lett. **57**, 768 (1993) [Pisma Zh. Eksp. Teor. Fiz. **57**, 755 (1993)];
    C. Broggini (MUNU Collaboration), in Proc. of *International Workshop on Neutrino Oscillations in Venice*, Venice, Italy, July 24–26, 2001, Nucl. Phys. B (Proc. Suppl.) **100**, 267 (2001) [hep-ex/0110026];
    TEXONO Collaboration, H.B. Li et al., Phys. Rev. Lett. **90**, 131802 (2003) [hep-ex/0212003].
89. J.F. Beacom and P. Vogel, Phys. Rev. Lett. **83**, 5222 (1999) [hep-ph/9907383].
90. A.S. Joshipura and S. Mohanty, Phys. Rev. D **66**, 012003 (2002) [hep-ph/0204305].
91. W. Grimus, M. Maltoni, T. Schwetz, M. Tórtola and J.W.F. Valle, Nucl. Phys. B **648**, 37 (2003) [hep-ph/02081326].
92. E. Meroni (BOREXINO Collaboration), in Proc. of *Europhysics Neutrino Oscillation Workshop*, Otranto, Italy, September 9–16, 2000, Nucl. Phys. B (Proc. Suppl.) **100**, 42 (2001).
93. T. Yanagida, in Proc. of the *Workshop on Unified Theories and Baryon Number in the Universe*, Tsukuba, Japan, 1979, eds. O. Sawada and A. Sugamoto (KEK report no. 79–18, Tsukuba, 1979);
    M. Gell-Mann, P. Ramond, and R. Slansky, in *Supergravity, Proceedings of the Workshop, Stony Brook, New York, 1979*, eds. F. van Nieuwenhuizen and D. Freedman (North Holland, Amsterdam, 1979).
94. R.N. Mohapatra and G. Senjanović, Phys. Rev. Lett. **44**, 912 (1980).
95. G. Altarelli and F. Feruglio, Phys. Rep. **320**, 295 (1999);
    H. Fritzsch and Z.-Z. Xing, Prog. Part. Nucl. Phys. **45**, 1 (2000) [hep-ph/9912358];
    S.M. Barr and I. Dorsner, Nucl. Phys. B **585**, 79 (2000) [0003058];
    G. Altarelli and F. Feruglio, hep-ph/0206077, to appear in *Neutrino Mass*, eds. G. Altarelli and K. Winter (Springer Tracts in Modern Physics);
    S.M. Barr, Talk given at the *NOON2001 Conference*, Tokyo, December 5–8, 2001, hep-ph/0206085;
    S.F. King, in Proc. of *XXth International Neutrino Conference on Neutrino Physics and Astrophysics*, Munich, May 25–30, 2002, Nucl. Phys. B (Proc. Suppl.) **118**, (2003) 267 [hep-ph/0208266];

C.H. Albright, Paper presented at the *Neutrinos and Implications for Physics Beyond the Standard Model Concerefence*, SUNY Stony Brook, October 11–13, 2002, hep-ph/0212090;

R.N. Mohapatra, Talk given at the *Xth International Workshop on Neutrino Telescopes*, Venice, March 11–24, 2003, hep-ph/0306016.

96. P.H. Chankowski and S. Pokorski, Int. J. Mod. Phys. A **17**, 575 (2002) [hep-ph/0110249].

97. W. Konetschny and W. Kummer, Phys. Lett. **70B**, 433 (1977).

98. A. Zee, Phys. Lett. **93B**, 389 (1980).

99. A. Zee, Phys. Lett. **161B**, 141 (1985).

100. L. Wolfenstein, Nucl. Phys. B **175**, 93 (1980).

101. C. Jarlskog, M. Matsuda, S. Skadhauge and M. Tanimoto, Phys. Lett. B **449**, 240 (1999) [hep-ph/9812282];

    P.H. Frampton and S.L. Glashow, Phys. Lett. B **449**, 240 (1999) [hep-ph/9906375].

102. K.R.S. Balaji, W. Grimus and T. Schwetz, Phys. Lett. B **508**, 301 (2001) [hep-ph/0104035].

103. P.H. Frampton, M.C. Oh and T. Yoshikawa, Phys. Rev. D **65**, 073014 (2002) [hep-ph/0110300];

    Y. Koide, in Proc. of *5th KEK Topical Conference: Frontiers in Flavor Physics (KEKTC5)*, Tsukuba, Ibaraki, Japan, November 20–22, 2001, Nucl. Phys. B (Proc. Suppl.) **111**, 294 (2002) [hep-ph/0201250].

104. K.S. Babu, Phys. Lett. B **203**, 132 (1988).

105. K.S. Babu and C. Macesanu, Phys. Rev. D **67**, 073010 (2003) [hep-ph/0212058].

106. G.B. Gelmini and M. Roncadelli, Phys. Lett. **99B**, 411 (1981).

107. J. Erler and P. Langacker, Phys. Lett. B **456**, 68 (1999) [hep-ph/9903476].

108. W. Grimus, R. Pfeiffer and T. Schwetz, Eur. Phys. J. C **13**, 125 (2000) [hep-ph/9905320].

109. E. Ma and Utpal Sarkar, Phys. Rev. Lett. **80**, 5716 (1998) [hep-ph/9802445].

110. J. Schechter and J.W.F. Valle, Phys. Rev. D **22**, 2227 (1980);

    S.M. Bilenky, J. Hošek and S.T. Petcov, Phys. Lett. **94B** 495 (1980);

    I.Yu. Kobzarev, B.V. Martemyanov, L.B. Okun and M.G. Shchepkin, Yad. Phys. **32**, 1590 (1980) [Sov. J. Nucl. Phys. **32**, 823 (1981)].

111. W. Grimus and H. Neufeld, Nucl. Phys. B **325**, 18 (1989).

112. W. Grimus and H. Neufeld, Phys. Lett. B **486**, 385 (2000) [hep-ph/9911465].

113. W. Grimus and L. Lavoura, Phys. Lett. B **546**, 86 (2002) [hep-ph/0207229].

114. M. Hirsch, M.A. Diaz, W. Porod, J.C. Romão and J.W.F. Valle, Phys. Rev. D **62**, 113008 (2000) [hep-ph/0004115];

    B. Mukhopadhyaya, Invited Contribution to a special issue of the Proc. of the Indian National Academy of Sciences, hep-ph/0301278;

    A. Bartl, M. Hirsch, T. Kernreiter, W. Porod and J.W.F. Valle, hep-ph/0306071.

115. W. Grimus and L. Lavoura, JHEP **07**, 045 (2001) [hep-ph/0105212];

    W. Grimus and L. Lavoura, in Proc. of *XXVth International School of Theoretical Physics*, Ustroń, Poland, September 10 – 16, 2001, Acta Phys. Polon. B **32**, 3719 (2001) [hep-ph/0110041].

116. W. Grimus and L. Lavoura, Phys. Rev. D **66**, 014016 (2002) [hep-ph/0204070].

117. W. Grimus and L. Lavoura, Eur. Phys. J. C **28**, 123 (2003) [hep-ph/0211334].

Lecture Notes in Physics

For information about Vols. 1–602
please contact your bookseller or Springer-Verlag
LNP Online archive: springerlink.com